SCHAUM'S OUTLINE OF

THEORY AND PROBLEMS

of

DESCRIPTIVE
GEOMETRY

●

BY

MINOR CLYDE HAWK

Chairman, Engineering Graphics Division
Carnegie Institute of Technology

●

McGraw-Hill
New York St. Louis San Francisco Auckland Bogotá
Caracas Lisbon London Madrid Mexico City Milan
Montreal New Delhi San Juan Singapore
Sydney Tokyo Toronto

27290

24 25 26 27 28 29 30 BAW 9 8 7 6 5 4 3

McGraw-Hill

A Division of The McGraw-Hill Companies

Preface

This book is designed primarily to supplement standard texts in introductory Descriptive Geometry, based on the author's firm conviction that numerous solved problems constitute one of the best means for clarifying and fixing in mind basic principles. Moreover, the statements of theory and principle are sufficiently complete that, with suitable handling of lecture-laboratory time, the book could be used as a text by itself.

In most engineering and technical schools the same course in Descriptive Geometry is offered to engineering and science students alike. They are given the same lectures and attend the same laboratory classes to solve the same types of problems. This book, therefore, attempts to present, as simply as possible, the basic principles which the author believes should be a vital and necessary part of every engineering and science student's education. A thorough understanding of these fundamental phases of graphical analysis should be sufficient to develop the student's potential for solving more difficult graphical problems to be encountered later on in individual fields of endeavor.

No attempt is made to reach every phase of work involved in such a wide field of study because, first of all, and unfortunately, in many engineering schools only a minimum of time is allocated for the study of Descriptive Geometry; secondly, the diversity of interests and departmental requirements make it imperative that only the fundamental items of interest to the majority of engineering and science students be included in a basic course of study.

The subject matter is divided into chapters covering duly-recognized areas of theory and study. Each chapter begins with statements of pertinent definitions, principles and theorems together with illustrative and descriptive material. This is followed by graded sets of solved and supplementary problems. The solved problems illustrate and amplify the theory, present methods of analysis, provide practical examples, and bring into sharp focus those fine points which enable the student to apply the basic principles correctly and confidently. Most of the practical problems are analyzed and solved step by step to insure complete understanding on the part of the student. It should be mentioned, however, that the solutions as given do not usually limit themselves to any one particular method. Most of the problems could be solved in several different ways, all consistent with proven principles of graphical analysis.

The author wishes to acknowledge the cooperation rendered by members of his staff whose assistance was invaluable. Professor H. L. McKee and Mr. Joseph Parris were especially helpful with problem selection. Others who graciously assisted with the typing of the manuscript and with the drawings are Mrs. Caroline Horey and Messrs. Roger Mohrlang, Henry Carmichael, Lynn Shaeffer and Chi Tsau. Special appreciation is expressed to Mr. Nicola Miracapillo, associate editor of the publishers, and to Mr. Henry Hayden, art editor, for valuable suggestions and fine spirit of cooperation. And finally, I extend heartfelt thanks to my wife and children for their grace and encouragement during the many months while the manuscript was being prepared.

M. C. Hawk

Carnegie Institute of Technology
August, 1962

CONTENTS

CONTENTS

CONTENTS

Chapter 1

Orthographic Drawing

1.1 INTRODUCTION

Orthographic drawing is the basis of all engineering drawing, and it is also the basis for the study of Descriptive Geometry. A well-trained engineer or technician must be able to pick up a drawing and understand it. This understanding, of necessity, involves the basic principles of orthographic drawing.

Generally speaking, a course in Engineering Drawing consists of drawing various objects in two or more views utilizing the principles of orthographic projection. These views may be projected on the three principal planes — horizontal, frontal and profile — or on auxiliary planes. In turn, the views may or may not be sectioned. Also included in a standard Engineering Drawing course would be problems dealing with pictorial drawing, freehand sketching, fasteners, piping drawings, working drawings, etc.

Many students entering an engineering school have had limited experience in orthographic drawing in the high school or technical school which may have prepared them for college. It may have only consisted of several weeks of Mechanical Drawing, but this previous contact with the principles involved in orthographic drawing forms a frame of reference which usually proves valuable in solving Engineering Drawing problems. Unfortunately, however, very few college students have been introduced to the basic principles of Descriptive Geometry before they enroll in the engineering school.

The question might then be asked, "Well, what is Descriptive Geometry?" Very briefly, Descriptive Geometry is the graphical solution of point, line and plane problems in space. These solutions are accomplished by means of the same principles of orthographic drawing which are involved in making a simple three-view drawing of an object. Therefore the student who seeks to understand Descriptive Geometry must also be acquainted with the basic principles of elementary Engineering Drawing. In other words, Descriptive Geometry is the graphical solution of the more advanced problems of Engineering Drawing; and both phases utilize the principles of orthographic drawing.

1.2 DEFINITIONS

The following terms are used repeatedly throughout this text, and therefore a thorough understanding of their meaning is imperative for the proper study of Descriptive Geometry.

(1) Orthographic Projection — the use of parallel lines of sight at 90° to an image plane.

(2) Image Plane — the plane which is perpendicular to the lines of sight. This plane is located between the eye of the observer and the object which is being viewed.

(3) Line of Sight — the path from the observer's eye to a particular point on the object. These lines of sight are parallel.

(4) Horizontal Plane — an image plane, all points of which are at the same elevation. [See Fig. 1-1(a) below.] The top, or plan view, is determined by the projection of the object on this plane. The lines of sight for this horizontal plane are vertical and are therefore perpendicular to it.

1

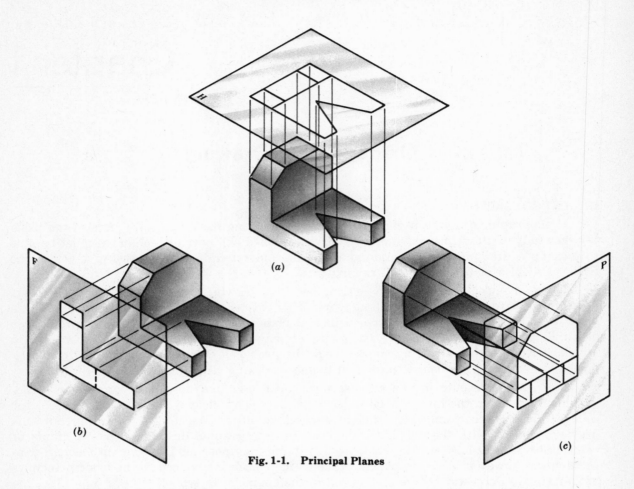

Fig. 1-1. Principal Planes

(5) *Frontal Plane* — an image plane at 90° to the horizontal and profile planes. [See Fig. 1-1(*b*) above.] The front elevation view is determined by the projection of the object on this plane. The lines of sight for this frontal plane are horizontal and are therefore perpendicular to it.

(6) *Profile Plane* — an image plane at right angles to both the horizontal and frontal planes. [See Fig. 1-1(*c*) above.] The right and left side elevation views are determined by the projection of the object on this plane. The lines of sight for this profile plane are horizontal and are therefore perpendicular to it.

(7) *Folding Line or Reference Plane Line* — the line made by the intersection of two image planes. It is designated as a long line, two short dashes and then another long line.

(8) *Elevation View* — any orthographic view for which the lines of sight are horizontal and perpendicular to the image plane. It may be projected from a plan view, other elevation views, or from inclined views. Any view projected from the plan view must be an elevation view.

(9) *Inclined View* — any orthographic view for which the lines of sight are neither horizontal nor vertical. It may be projected from an elevation view or other inclined views, but never from a plan view.

(10) *Line* — the path of a moving point.

(11) *Straight Line* — the path of a moving point proceeding constantly in the same direction. A line having a definite length is determined by its extremities. However, any two points on the line may be chosen for the purpose of locating the entire line in another view. The end view of a line is a point which represents all points on the line.

(*12*) *Level Line* — a line which is parallel to the horizontal image plane and which therefore has all points on the line at the same elevation. It will appear in its true length in the plan view. [See Fig. 1-2(*a*) below.]

(*13*) *Frontal Line* — a line which lies parallel to the frontal image plane. The line must show in its true length in the front view even though it may be level, vertical, or inclined. [See Fig. 1-2(*b*) below.]

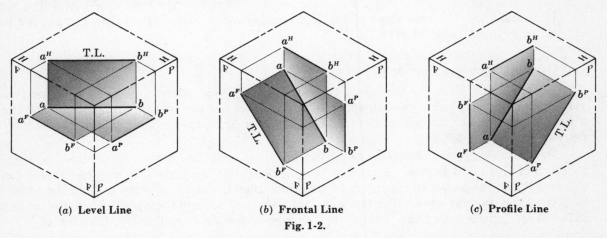

(*a*) Level Line	(*b*) Frontal Line	(*c*) Profile Line

Fig. 1-2.

(*14*) *Profile Line* — an inclined line which lies parallel to the profile image plane. The line must show in its true length in the profile view. [See Fig. 1-2(*c*) above.]

(*15*) *Vertical Line* — a line which is perpendicular to a level plane. It will appear in its true length in any elevation view.

(*16*) *Inclined Line* — a line neither vertical nor horizontal but which may appear in its true length in either the frontal or profile planes. It cannot appear in its true length in the plan view.

(*17*) *Oblique Line* — a line inclined to all three principal planes. It cannot appear in its true length in any of the three principal planes.

(*18*) *Contour Line* — a straight or curved line used on topographical drawings which locates a series of points at the same elevation. Therefore a contour line is a level line. [See Fig. 1-3]

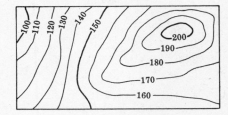

Fig. 1-3.   Contour Lines

(*19*) *Bearing* — the angle between the plan view of a line and a line running due north and south. North is always assumed to be at the top of the drawing sheet unless otherwise indicated by a directional arrow on the drawing. The acute angle is usually given for the bearing. In Fig. 1-4(*a*) the line *AB* has a bearing of S 45° E. This means that the line is located 45° east of the south line. If the bearing is taken from *B* to *A*, then it would be expressed as N 45° W. The bearing can only be found on the plan view, and furthermore, it is not affected in any way by the fact that the line may be level or inclined. Fig. 1-4(*b*) shows some sample bearings.

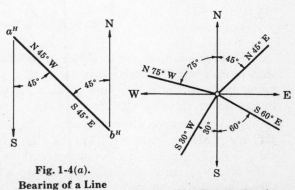

Fig. 1-4(*a*).
Bearing of a Line

Fig. 1-4(*b*).   Sample Bearings

(20) Normal View of a Line or Plane — the view which shows the true length of the line or the true size of the plane. A normal view of a plane shows the true size of any angle on the plane and the true length of any line which lies on the plane.

(21) Slope of a Line — tangent of the angle that the line makes with a horizontal plane. Two conditions must be met in order to determine the slope of a line. *First,* the line must be shown in an elevation view; *secondly,* the line must appear in its true length in this elevation view. *Note:* An inclined view may show the true length of a line but it cannot show the true slope of the line because a horizontal plane does not appear as an edge in an inclined view.

1.3 MULTIVIEW DRAWINGS

By multiview drawing we mean a logical arrangement of two or more orthographic views of an object, shown on a one-plane sheet of drawing paper. The relationship of the views are dependent upon the fact that both views are projected on image planes perpendicular to each other.

Before an engineer can draw an object in space, whether it be a line, plane, or combination of lines and planes, he must be able to visualize the object. Once he establishes the position of the object in his mind, he then imagines that he is moving around the object to secure the various views necessary to complete his understanding of what the object looks like. This is called the "direct" or "change-of-position" method of drawing.

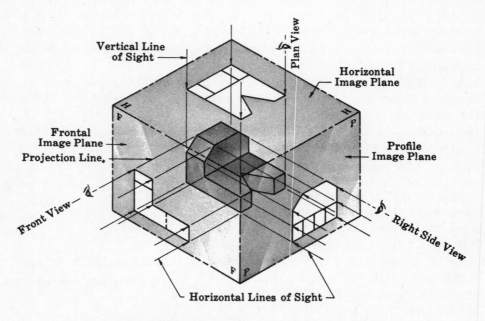

Fig. 1-5. The Three Principal Projection Planes

In Fig. 1-5, in order to obtain a right side view of the block, the observer must imagine that he has left the position in front of the block and is now looking at the block from the right side position. His line of sight is perpendicular to the profile image plane and, as in the front view, is still horizontal.

If the observer views the object from any position as he moves around the object and still maintains a horizontal line of sight, he will be obtaining an elevation view since the elevation of all points on the object will maintain their same relationship. The front, rear, left side, and right side views are all elevation views. Likewise, any views in between these four positions which have horizontal lines of sight are called auxiliary elevation views.

In Fig. 1-5 above, we have an object which lies behind the frontal plane, to the left of the profile plane, and below the horizontal plane. This position in space is referred to as the third-angle projection, which forms the basis for practically all engineering drawing in the United States. In the third-angle projection system, the image plane is imagined to be located between the observer and the object. In third-angle projection, the block shown in Fig. 1-5 would have the six basic views as shown in Fig. 1-6.

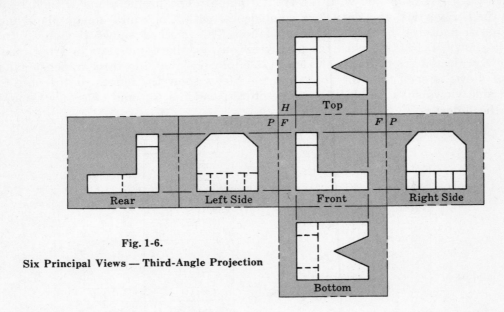

Fig. 1-6.

Six Principal Views — Third-Angle Projection

In most foreign countries, as well as for some architectural and structural drawings in the United States, the first-angle projection is employed. In the first-angle projection, the object is imagined to be located between the observer and the image plane. Fig. 1-7 shows the six basic views of our block as they would be drawn using the first-angle projection system.

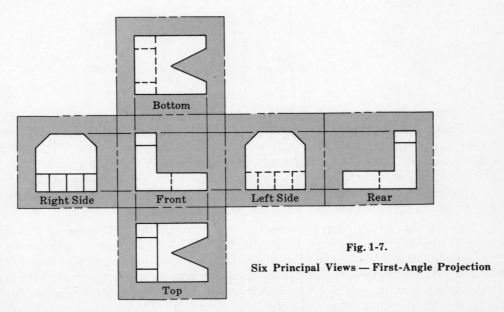

Fig. 1-7.

Six Principal Views — First-Angle Projection

Now, since most orthographic drawings of an object require three views, and since these views are mutually perpendicular to each other, it becomes necessary to revolve two of the image planes until they lie in the plane of the third image plane. These three image planes can now be located in their relative positions on a one-plane sheet of drawing paper. The method commonly practiced in a course of Engineering Drawing is that of maintaining the position of the frontal plane and revolving the horizontal plane 90° about the *H-F* horizontal axis until it falls in line with the frontal plane. The profile plane is then revolved 90° about the *F-P* vertical axis until it is coincidental with the frontal image plane and the revolved position of the horizontal image plane. This method shows the object in the position as shown in Fig. 1-8 with the plan view on top, the front view in front, and the profile view projected from the front view. It is obvious that the three-dimensional relationship of views is such that the front and top views show the length of the object; the front and side views show the height of the object; and the top and side views show the depth of the object.

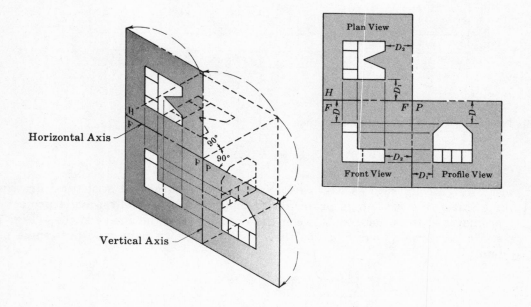

**Fig. 1-8. Revolving of Horizontal and Profile Planes
to Coincide with Frontal Plane**

In Fig. 1-8, it should be noted that the distance from the top of the block to the horizontal image plane is the same in both the front and profile views. This would also be true for any elevation view that is directly related to the plan view. The distance between the block and the frontal image plane is the same in both the plan and profile views. These facts are very important, and they form the basis for the transfer of distances from one view to another. The distance from the block to the profile image plane is the same in both the plan and front views.

An alternate to the above method of revolving planes is that of maintaining the position of the plan view and then revolving both the frontal and profile image planes until they lie in the same plane as the horizontal image plane, or plan view. Fig. 1-9 below shows this alternate method which locates the plan view on top, the front view in front, and the profile view is now projected from the plan view. Both methods are correct and the student should feel free to use either method to solve the problems unless the given data makes more practical the use of one method in preference to the other.

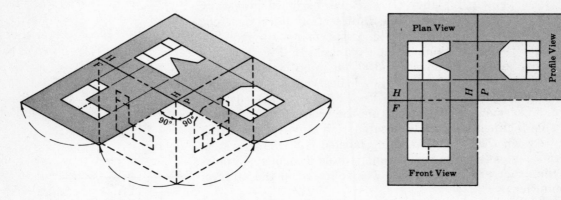

Fig. 1-9. Revolving of Frontal and Profile Planes to Coincide with Horizontal Plane

1.4 NOTATION

Since, in most instances, this book is being used as a supplementary text, it is quite likely that the notation system used by the author is different from that used by the student in his course of Descriptive Geometry. Fig. 1-10 shows several notation methods which are recommended by authors of Descriptive Geometry textbooks. However, even though the notation systems may vary, the methods of problem solving are all based on the same basic principles.

As noticed in Figures 1-8 and 1-9, the intersection of the horizontal and frontal image planes is designated by placing the letter H on the side which shows the plan view, and the letter F is placed on the side which gives us the front view. Similarly, the intersection of the plan and profile views is designated by placing the letter H on the plan side of the reference line and the letter P is located on the profile side of the intersection. If the profile view is projected from the front view, then the letter F is placed on the frontal plane side of the intersection line, and the letter P would be located on the profile side.

Since most Descriptive Geometry problems require views which are projected on planes other than

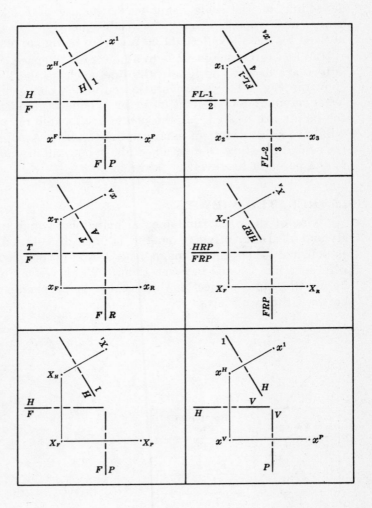

Fig. 1-10.

Various Systems of Notation

the three principal planes (H, F, P), it becomes necessary to label these additional plane intersection lines by some logical system. The author suggests using numbers to designate these additional image planes. In Fig. 1-11, we have this system demonstrated. The auxiliary elevation view projected from the plan view is designated view 1, and the folding line between the two views is labeled H-1. An inclined view is projected from the front elevation view, and this inclined view is designated view 2 with the folding line between the two views being labeled F-2. This system offers a logical system for anyone who is desirous of knowing the exact procedure which was followed in the attempt at solution.

The reference plane lines, or folding lines, are represented by a long line broken by two short dashes. This line should be a dark line, yet fine enough to insure accuracy when stepping off distances from it. On some problems, depending on the scale being used, the use of thick lines could cause repeated error and, subsequently, the answer derived by the student could be far from being correct.

When reference is made to a line itself in space, capital letters are used to designate the line, such as the line AB

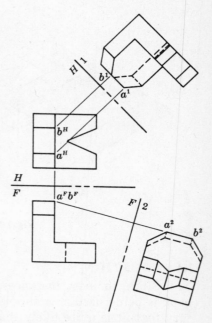

Fig. 1-11.   Notation of Line AB

shown in Fig. 1-11. However, the individual points on a view are labelled with a lower case letter having a superscript letter or exponent number corresponding to the view on which the points are located. If the points are located on any of the three principal planes, they will have superscripts of small capital letters. A point which has been revolved will follow the same procedure as mentioned above but will also have a lower case letter r as a subscript to the point in its revolved position.

1.5   RELATED VIEWS

One of the most fundamental principles which must be thoroughly understood by the student of Descriptive Geometry is that of relationship of views. This relationship is established when their image planes are perpendicular to each other and they have a reference line located between them. When views are related, the two views of any point lie on the same projecting line which is perpendicular to the reference line which lies between the two views.

Views	Related to View
$P, F, 1, 3$	H (Plan)
$H, 2$	F (Front)
$H, 4$	3 (Aux. Elev.)
$3, 5$	4 (Inclined)

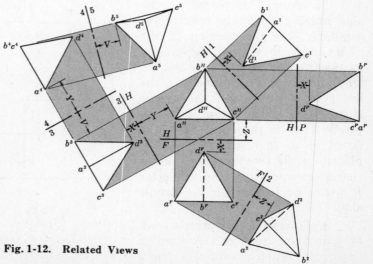

Fig. 1-12.   Related Views

In Fig. 1-12 above, we have several views of a simple object. You will notice that the location of points for "every other view" is the same distance from the folding line. Distance X in the profile view is the same as distance X in the front elevation and auxiliary elevation views. The distance Y in the inclined view 4 is the same as that in the plan view. The distance Z in the inclined view 2 is the same as the distance Z in the plan view. Therefore it may be stated with the force of a rule that in all views which are related to a common view, any point on an object is the same distance away from the folding line. It will also be noted in Fig. 1-12 that every view adjacent to the plan view shows the true height of the object. The table shows which views are related to each other.

1.6 VISUALIZATION

Many students of engineering and science are able to follow certain rules and patterns, but when they are called upon to visualize the item they are at a loss. This is usually the result of improper visualization study of basic principles involving the lines and planes.

LINES. As far as direction is concerned, straight lines may be classified as vertical, horizontal, or inclined. Fig. 1-13 shows several positions of lines in space. It is suggested that the student try to visualize the various positions of the line AB by holding a pencil to indicate the directions as shown. At (1) we have shown an inclined frontal line which appears in its true length in the front elevation view. At (2) the horizontal line appears in its true length in the plan view. The oblique line at (3) does not appear in its true length in any of the principal views. At (4) the vertical line will appear as a point in the plan view, and in its true length in both the front elevation and profile views. The profile line at (5) will appear in its true length in the profile view. In both (6) and (7) the oblique line does not appear in its true length in any of the three given views. At (8) the horizontal-frontal line appears in its true length in both the plan and front elevation views. The horizontal line at (9) appears as a point in the front view and in its true length in both the plan and profile views.

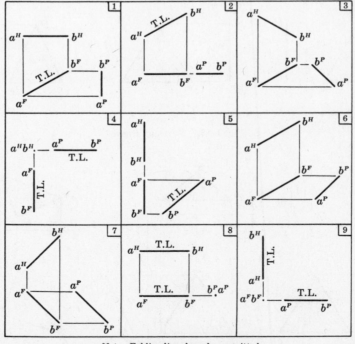

Note: Folding lines have been omitted

Fig. 1-13   Visualization of Lines

PLANES. As far as visualization is concerned, planes may be classified as horizontal, vertical, or inclined. It should be made clear that every plane surface must appear either as an edge or as a plane of similar configuration. In other words, a triangular surface will always appear as an edge (a line) or as a triangle. Likewise, a square-shaped plane must always appear as an edge or as a four-sided plane. This four-sided plane will either be a square, a rectangle, or a parallelogram.

In Fig. 1-14 we have three principal views of a horizontal plane. The plane appears as an edge in both the front and profile views, or any other elevation view, whereas the true size of any horizontal plane will appear in the plan view.

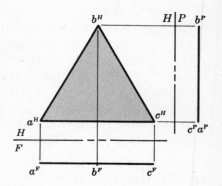

Fig. 1-14. **Horizontal Plane**

Fig. 1-15 below shows three possible positions for a vertical plane. At (*a*) we have a vertical plane which is parallel to the frontal image plane; it is therefore in its true size in the front view. It is represented as an edge in both the plan and profile views. The vertical plane at (*b*) is shown parallel to the profile image plane, thus being represented as an edge in both the plan and front views. The true size of the plane appears in the profile view. At (*c*) the vertical plane is inclined to both the frontal and profile image planes; therefore the plan view shows the plane as an edge and the true size of the plane is not shown in the given views.

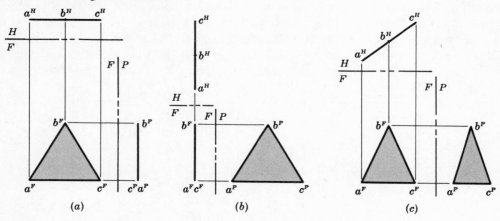

Fig. 1-15.   **Vertical Planes**

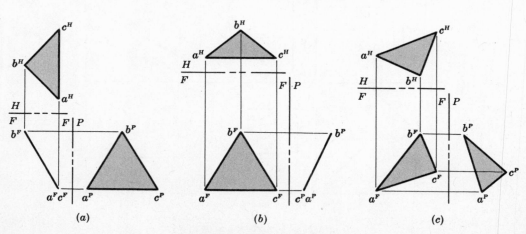

Fig. 1-16.   **Inclined Planes**

In Fig. 1-16 above we have some typical positions of an inclined plane. An inclined plane is one which is neither vertical nor horizontal. It cannot appear in its true size in any of the three basic views — plan, front, or profile. It may appear as an edge in the front view (a), or in the profile view (b), but it cannot appear as an edge in the plan view. At (c) we have the inclined plane in its most common position, that of being inclined to all three principal planes of projection. In this case, it will not appear as an edge in any of the three principal views. It is usually referred to as an oblique plane.

1.7 PROBLEM LAYOUT

Textbooks in Descriptive Geometry use various methods to express given data for the problems which usually accompany the text. Some show the given data on a sheet ready for further development or a solution. See Fig. 1-17(a) below. Others express the data in statement form requiring the student to plan for proper spacing and locating given points. The scale is usually established by the instructor beforehand. In this system the relationship between points is given by direction and elevation. Thus in Fig. 1-17(b) line AB bears N 30° E, and B is 100′ map distance from A and 40′ below A. At (c) the following information is given about plane ABC: B is 20′ east, 30′ north of A and 10′ above A; point C is 40′ east, 10′ south of A and 15′ below A. Having been given a suitable scale by the instructor, the student should locate point A at a convenient position on the paper and then step off the distance 20′ east and 30′ north of A; this establishes point B in the plan view. Point B in the front view would be projected directly below its location in the plan view and at an elevation of 10′ above point A, which has also been projected directly from the plan view. Point C would be located in a similar manner.

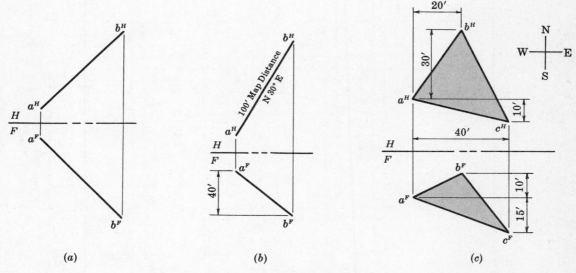

(a)   (b)   (c)

Fig. 1-17. Problem Layout

Still another method of problem layout which is used extensively is that of coordinate dimensions. Every point has three coordinates in space even though in many problems one or more of these coordinates may be omitted in the given data. If a coordinate is unknown and the complete location of the point is part of the problem, the letter X is inserted in the given data. Example: Point $A(3, 4\frac{1}{2}, X)$.

Coordinate dimensions are usually given in inches and the drawing is made full scale, regardless of the fact that the scale of the problem is specified otherwise. Coordinate paper, $8\frac{1}{2}″ \times 11″$, divided into quarters of an inch is very convenient for problems using coordinate dimensions.

The "origin", or zero point, is assumed to be the lower left hand corner of the drawing space.   Example of coordinate dimensioning for the plane ABC established by the line AB and the point C: $A(2, 3, 7)$, $B(4, 3\frac{1}{2}, 6)$, $C(5, 2, 7\frac{1}{2})$.   See Fig. 1-18 below.   For locating point A, the first ordinate 2 tells us that the plan and front views of A are located 2″ from the left border line.   The second ordinate 3 locates the front view of A, 3″ above the horizontal base line which passes through the origin.   The third ordinate 7 establishes the point A in the plan view 7″ above the horizontal base line.   Points B and C would be located in a similar manner.

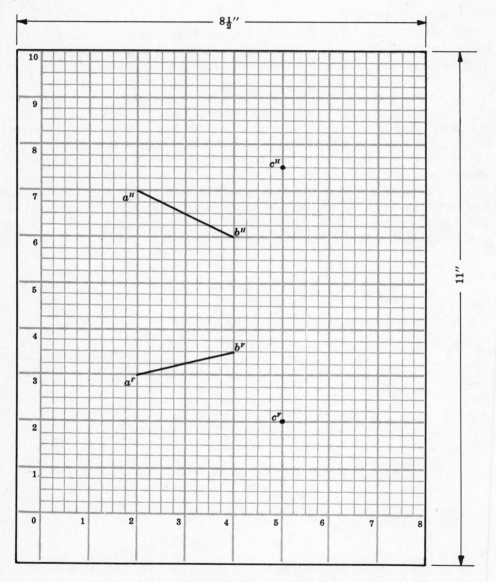

Fig. 1-18.   Problem Layout — Coordinate Data

Chance of subsequent error and time-saving are two good reasons for the student to label each point as soon as it is located.   It is also advisable for the student to check the location of points with the given data.   A common error is that of confusing the east and west directions.   North, of course, is always assumed to be at the top of the drawing sheet.

The answers to the problems in this text have been derived by graphical solution, and in most cases they have not been substantiated by subsequent mathematical calculations.

In correcting Descriptive Geometry problems, the instructor should allow a "tolerance" commensurate with the scale being used. The scales used for the solved problems in this text are such that the student can directly check the drawings. In practice the scales should be altered, if necessary, to suit the size of the drawing paper being used.

Some of the supplementary problems in the text are based on the use of an Architect's scale, whereas others require the use of an Engineer's scale. It is suggested that the student purchase both types of scales in order to insure a greater degree of accuracy as well as to be acquainted with the use of both types.

Solved Problems

1. **Given:** Plan and front elevation views in Fig. 1-19.
   Problem: Draw profile views from both the plan and front elevation view.
   Solution: See Fig. 1-19 below.

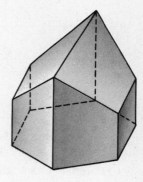

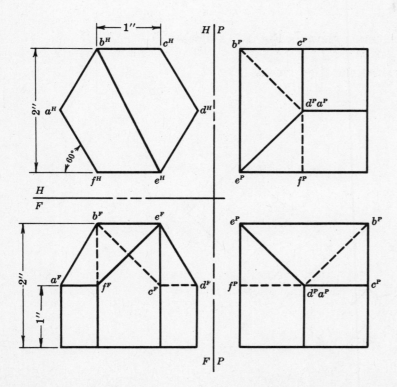

Fig. 1-19

2. **Given:** Plan and front elevation views in Fig. 1-20. Location of *H*-1 and *F-P*.
   Problem: Draw an auxiliary elevation view off the plan, and a profile view from the front elevation view.

   Solution: See Fig. 1-20.

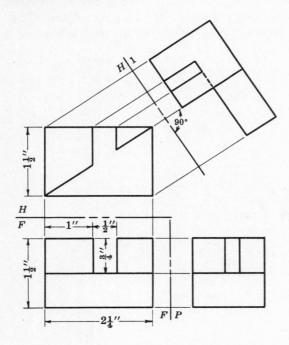

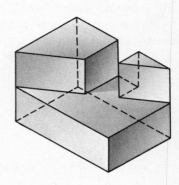

Fig. 1-20

3. **Given:** Plan and front elevation views in Fig. 1-21. Location of *H*-1 and 1-2.
   Problem: Draw the required elevation views.

   Solution: See Fig. 1-21.

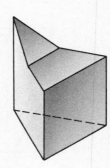

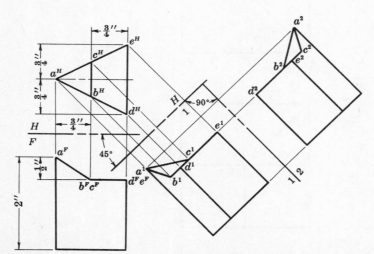

Fig. 1-21

4. **Given:** Plan and front elevation views in Fig. 1-22. Location of *H-P* and *F*-1.
    Problem: Draw a profile view projected from the plan, and an inclined view projected
              from the front view.

    Solution: See Fig. 1-22.

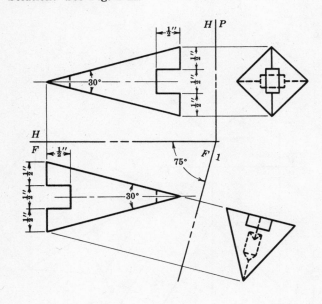

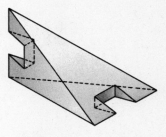

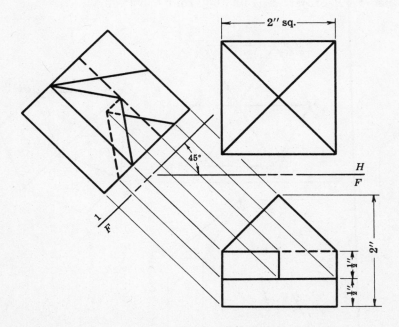

Fig. 1-22

5. **Given:** Plan and front elevation views in Fig. 1-23. Location of *F*-1.
    Problem: Draw an inclined view projected from the front elevation view.

    Solution: See Fig. 1-23.

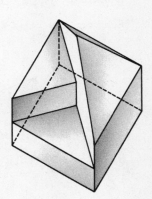

Fig. 1-23

6. Given: Plan and front elevation views in Fig. 1-24.   Location of *F*-1.
Problem: Draw an inclined view projected from the front elevation view.

Solution: See Fig. 1-24.

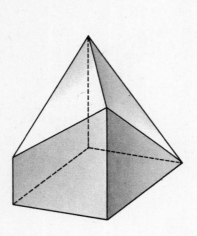

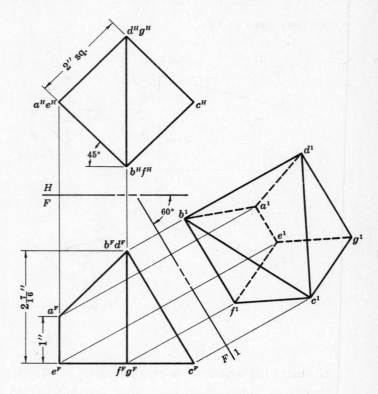

Fig. 1-24

7. Given: Plan and front elevation views in Fig. 1-25.   Location of *F*-1.
Problem: Draw an inclined view projected from the front elevation view.

Solution: See Fig. 1-25.

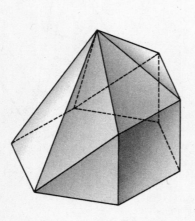

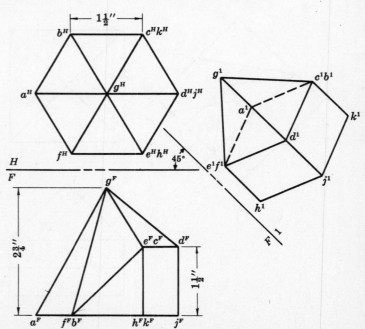

Fig. 1-25

8. **Given:** Plan and front elevation views in Fig. 1-26. Location of *H-P*, *P*-1, and *F*-2.

Problem: Draw the required profile and inclined views.

Solution: See Fig. 1-26.

9. **Given:** Plan and front elevation views in Fig. 1-27. Location of *H*-1.

Problem: Draw an auxiliary elevation view as indicated.

Solution: See Fig. 1-27.

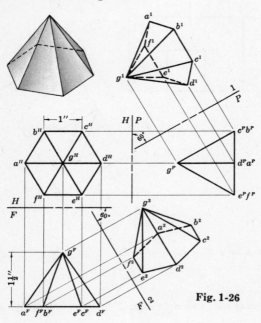

Fig. 1-26

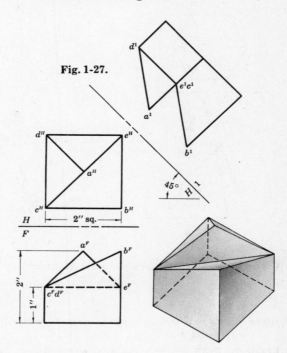

Fig. 1-27.

10. **Given:** Plan and front elevation views in Fig. 1-28. Location of *F-P*, *H*-1, and 1-2.

Problem: Draw the required profile, auxiliary elevation, and inclined views.

Solution: See Fig. 1-28.

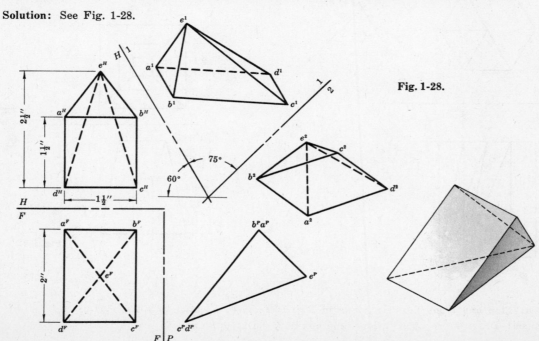

Fig. 1-28.

Supplementary Problems

11. **Given:** Front view and partial plan view as shown in Fig. 1-29 below of a truncated pyramid having a regular hexagonal base.
   Problem: Complete the plan view and draw an inclined view showing the true size of the cut surface.

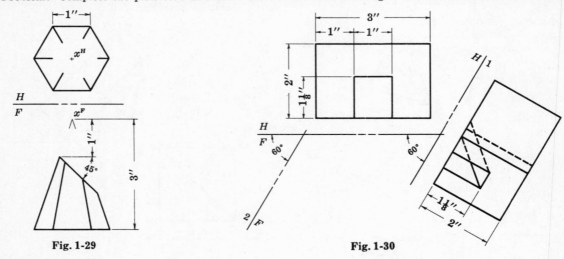

Fig. 1-29   **Fig. 1-30**

12. **Given:** Plan and auxiliary elevation views in Fig. 1-30 above. Location of F-2.
   Problem: Draw a front elevation view and an inclined view as indicated.

13. **Given:** Plan and front elevation views of a regular hexagonal prism in Fig. 1-31 below. Location of H-1 and 1-2.
   Problem: Draw the required auxiliary elevation and inclined views.

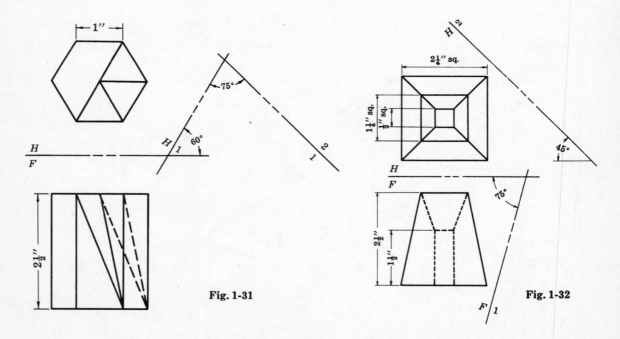

Fig. 1-31   **Fig. 1-32**

14. **Given:** Plan and front elevation views in Fig. 1-32 above. Location of F-1 and H-2.
   Problem: Draw the required inclined and auxiliary elevation views.

15. **Given:** Plan and front elevation views in the adjacent Fig. 1-33. Location of F-1 and H-2.

　　Problem: Draw the required auxiliary elevation and inclined views.

16. **Given:** Front elevation and profile views in Fig. 1-34 below. Location of F-1 and F-2.

　　Problem: Draw the required inclined views.

17. **Given:** Plan and front elevation views in Fig. 1-35 below. Location of H-1 and F-P.

　　Problem: Draw the profile and auxiliary elevation views as required.

18. **Given:** Plan and front elevation views as shown in Fig. 1-36 below of a truncated pyramid having a regular hexagonal base. Location of F-1 and 1-2.

　　Problem: Draw the required inclined views.

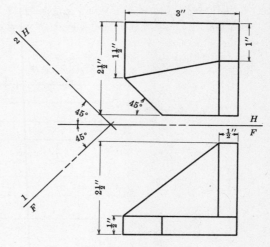

Fig. 1-33

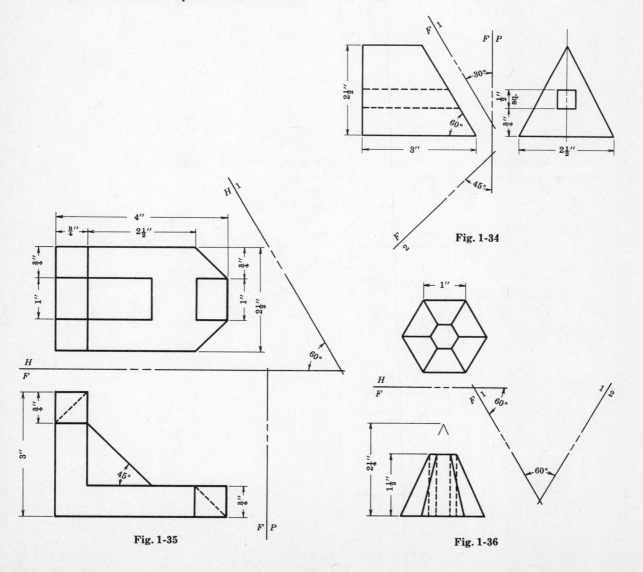

Fig. 1-34

Fig. 1-35

Fig. 1-36

19. Given: Plan and profile views in Fig. 1-37 below. Incomplete front elevation view. Location of *F*-1.
Problem: Complete the front elevation view and draw inclined view 1.

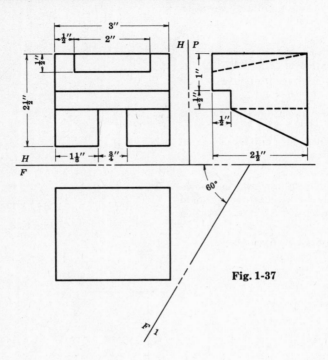

Fig. 1-37

20. Given: Plan and front elevation views in Fig. 1-38 below. Location of *H*-1, 1-2, and *F*-3.
Problem: Draw the required elevation and inclined views.

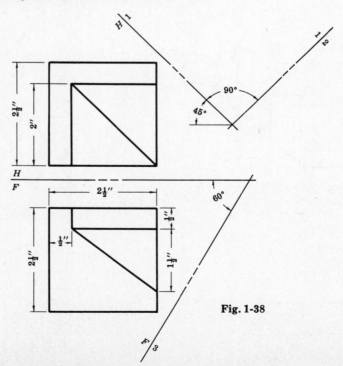

Fig. 1-38

Chapter 2

Fundamental Views –
Point, Line, and Plane

The projecting of a point or line from one view to another is vitally important to the solution of Descriptive Geometry problems. If a point in space is located on a line in one view, it must appear on that line in every view of the line.

If we have an oblique line in space, which contains point X in one view, as shown in Fig. 2-1 below, the point's location in the other two principal views would be found by simple projection from one view to another.

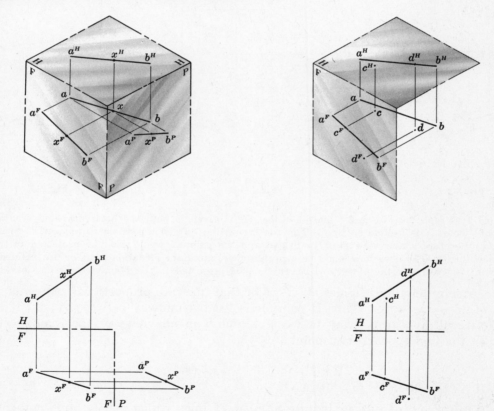

Fig. 2-1   Fig. 2-2

In Fig. 2-2 above, the point C in the front view appears to be on the line AB, but we notice in the plan view that point C is actually located in front of the line. Likewise, point D in the plan view appears to be located on the line AB but, again, we notice that the elevation of point D is actually lower than any point on the line. Therefore a point which appears to be located on a line in one view may not be on the line at all but may lie in front of the line, behind it, above it, or below it.

2.1 To PROJECT a POINT on a LINE from ONE VIEW to ANOTHER VIEW when the LINE is PARALLEL to a PRINCIPAL PLANE

Analysis: Project the point at 90° to the folding line between the two views.

Example 1: *Frontal Line* (Fig. 2-3 below). Line AB is given in both the horizontal and frontal projection planes. Point X is located on line AB in the plan view. To locate its position in the frontal plane, simply project at 90° to the H-F folding line and obtain x^F. Folding line F-P is placed perpendicular to the H-F line. Points A, B, and X in the profile view will be the same distance away from the frontal plane as they are in the plan view. The line will appear parallel to the frontal plane in both the horizontal and profile projection planes.

Example 2: *Level Line* (Fig. 2-4 below). Line AB is given in both the horizontal and frontal projection planes. Point X is located on line AB in the front view. To obtain its position in the plan view of line AB, project perpendicular to the H-F folding line to obtain x^H. Place folding line H-P perpendicular to the H-F line and project points A, B, and X from the plan view to the profile view. The distance these points will appear from the horizontal plane will be the same in both the front and profile views. The line will appear parallel to the horizontal plane in both the frontal and profile projection planes.

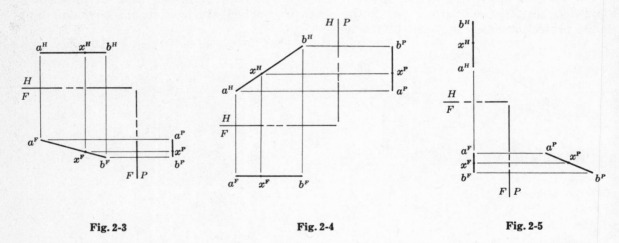

Fig. 2-3   **Fig. 2-4**   **Fig. 2-5**

Example 3: *Profile Line* (Fig. 2-5 above). Line AB is given in both the horizontal and frontal projection planes. Point X is located on line AB in the front view. It is impossible to project it directly to the plan view, therefore a new view must be drawn and the point X would then be projected to the line AB in this new view. This new view could be a profile view, auxiliary elevation view, or an inclined view. After the point is located in the new view, it can then be projected to the plan view or any other view.

Note: A point on a line divides all views of that line into proportional segments which always have the same ratio. Thus if point X is located midway between A and B in the front view, it will appear midway between A and B in all views of the line, except, of course, when the line appears as a point.

2.2 To LOCATE a POINT in a PLANE

Analysis: Choose any one of an infinite number of lines which can be drawn through a point on a plane. The various views of the given point must lie on the corresponding views of the chosen line passing through the point.

Example: The given data reveals plane ABC containing the point X in the plan view. See Fig. 2-6 below. Draw a line, YZ, through the point X on the plan view of the plane. Locate the line in the front elevation view by simple orthographic projection. Be sure that the points Y and Z are located on lines AC and BC respectively.* Since line YZ passes through point X in the plan view, project point X down to the front elevation view and its location will be on the line YZ in the front view.

Note: The same principle would apply for the projection of any line on a plane from one view to another.

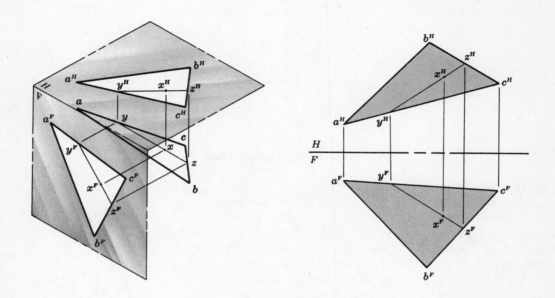

Fig. 2-6. To Locate a Point in a Plane

2.3 FUNDAMENTAL VIEWS

Practically every problem which must be solved graphically by an engineer or draftsman will involve certain basic or fundamental views. From the understanding of four basic views is derived the solutions of the more difficult space problems involving points, lines, and planes.

A thorough knowledge of these four fundamental views and related material will form a solid foundation for developing the student's potential to analyze and solve all problems involving Descriptive Geometry. For this reason, the problems at the end of this chapter are both numerous and varied.

The four fundamental views are:
1. The true length of a straight line
2. The straight line appearing as a point
3. The plane appearing as an edge
4. The plane appearing in its true size and shape

2.3.1-A The TRUE LENGTH of a STRAIGHT LINE

Analysis: Any line in space which is parallel to an image plane will be projected on that plane in its true length. Similarly, if a line is parallel to a folding line (reference line) in one view, it will appear in its true length in the related view on the other side of the folding line.

Examples: In Fig. 2-7 below we have several examples of lines which appear in their true length in the three principal views. At (*a*) the line is vertical and therefore appears in its true length in any elevation view. At (*b*) and (*c*) we have shown level lines which appear as true length in the plan view. At (*d*) is shown a frontal line which reveals its true length in the front view. And finally, at (*e*) is illustrated a profile line which shows its true length in the profile view.

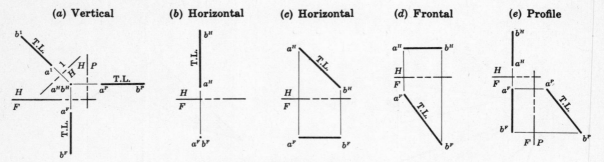

(a) Vertical   *(b)* Horizontal   *(c)* Horizontal   *(d)* Frontal   *(e)* Profile

Fig. 2-7. True Length Lines

In the case of an oblique line which does not appear in its true length in any of the principal views, a new view is drawn by placing an image plane parallel to the line as shown in one of the given views. In Fig. 2-8, for example, the line *AB* is given in both the plan and front elevation views. It is obvious from observation that the line is neither parallel nor perpendicular to any of the three principal planes; therefore it is an oblique line.

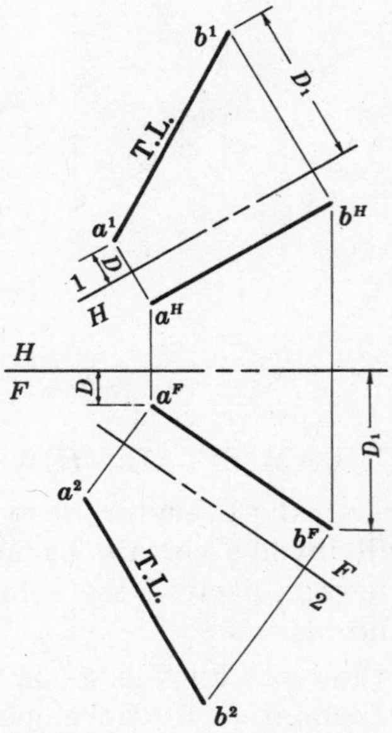

Since a line must be parallel to an image plane in order to be shown in its true length, a new image plane is located parallel to the plan view of the line. The intersection of this auxiliary elevation image plane with the horizontal image plane is designated as *H*-1 folding line. This new view, being an elevation view projected directly from the plan, will have the same elevational relationship between its points as the front view. In other words, points *A* and *B* in this auxiliary elevation view will be the same distance from folding line *H*-1 as they are from *H*-*F* to their corresponding points in the front elevation view.

Fig. 2-8. True Length of an Oblique Line.

Another means of determining the true length of the line is that of projecting an inclined view directly from the front elevation view. By passing an inclined image plane parallel to the front view of the line and projecting perpendicular to it, we obtain the true length of the line in view 2.

The folding line *H*-1 can be placed on either side of the plan view of line *AB* and parallel to it without affecting the true length of the line. Likewise, folding line *F*-2 can be placed parallel to, and at any distance from, the front view of line *AB*. Available drawing space and clarity should determine the location of the folding line. **If possible,** the student should avoid "overlapping" views.

The student may be called upon to check his graphical solution for finding the true length of a line by also making a mathematical calculation. The graphical means of determining the true length of a line is usually accurate enough for most requirements. However, given sufficient data, a mathematical solution is more accurate and easily calculated.

In Fig. 2-9(*a*) below we have shown an oblique line *AB* which represents the diagonal of an imaginary rectangular box. If the length of the box is 2″, its depth ¾″, and its height 1″, the true length of the line would be equal to the square root of the sum of the squares of *L, D,* and *H.*

$$\text{True length of } AB = \sqrt{L^2 + D^2 + H^2} = \sqrt{2^2 + (\tfrac{3}{4})^2 + 1^2} = \sqrt{4 + .5625 + 1}$$
$$= \sqrt{5.5625} = 2.3585 \text{ in.}$$

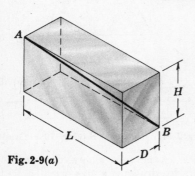

Fig. 2-9(a)

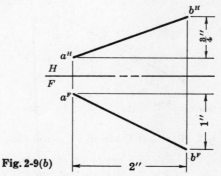

Fig. 2-9(b)

In Fig. 2-9(*b*) we have shown the plan and front elevation views of this oblique line.

2.3.1-*B* The TRUE SLOPE of a STRAIGHT LINE

Analysis: The true slope of a line is the tangent of the angle that the line makes with a horizontal plane. The angle between the true length of the line and a horizontal plane is called the slope angle. The slope angle of any line can be seen only in the elevation view which shows the line in its true length.

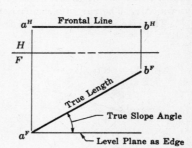

Fig. 2-10.

Note: Even though an inclined view may show the true length of a line, it cannot show the true slope because an inclined view cannot show a horizontal plane as an edge.

Examples: In Fig. 2-10, we have shown the true slope angle for a frontal line. Fig. 2-11 shows an oblique line *AB* which must first be shown in its true length in an elevation view before its slope can be determined. As noted above, the inclined view 2 may show the true length of the line but it cannot show the slope.

The angle of slope is usually expressed in degrees. However, in civil engineering the slope is expressed either as per cent grade (Fig. 2-12 below), or as batter (Fig. 2-13 below).

The % grade is given by the expression

$$\% \text{ grade} = \frac{\text{Vertical rise}}{\text{Horizontal run}} \times 100$$

Perhaps the most convenient method of measuring the horizontal run is to use an engineer's scale since 10 main divisions on each of its various scales contain 100 subdivisions.

Fig. 2-14 below shows the method of indicating the % grade of a highway.

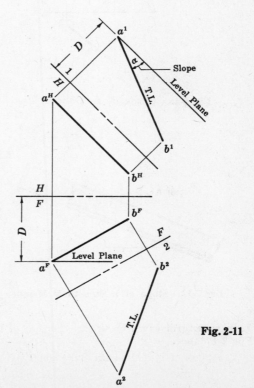

Fig. 2-11

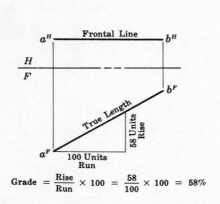

$$\text{Grade} = \frac{\text{Rise}}{\text{Run}} \times 100 = \frac{58}{100} \times 100 = 58\%$$

Fig. 2-12. Per Cent Grade of a Frontal Line

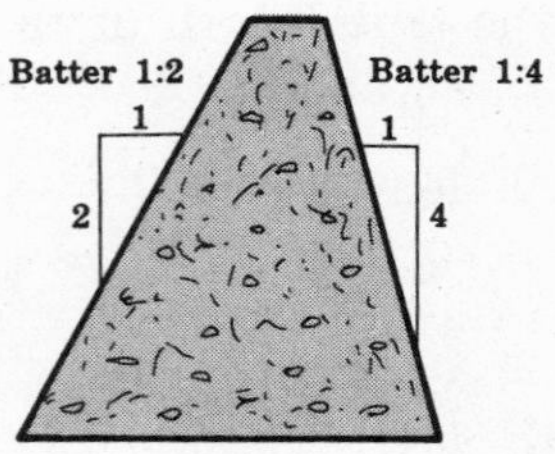

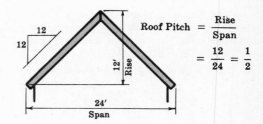

Fig. 2-13. Batter

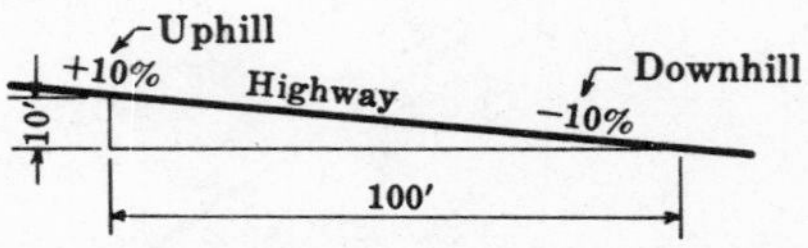

Fig. 2-14. Highway Having 10% Grade

To find the per cent grade of any oblique line, the line must be first shown true length in an elevation view. In Fig. 2-15 below the grade calculation is made by drawing a line parallel to the *H*-1 folding line and setting off 100 units along this line. The rise is measured perpendicular to the *H*-1 line, thus establishing the 58% grade.

Fig. 2-16 below illustrates the method for obtaining the grade of an oblique line from the plan and front views only. The run is measured on the plan view of the line by setting off 100 units and establishing point x^H. The rise will appear in the front view as the difference in elevation between a^F and x^F.

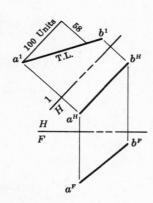

Fig. 2-15. Grade of an Oblique Line

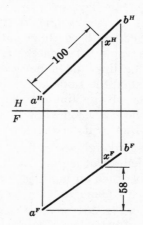

Fig. 2-16. Grade Measured from Plan and Front Views Only

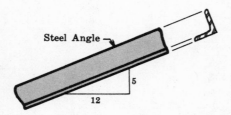

Fig. 2-17. Slope of a Structural Member

$$\text{Roof Pitch} = \frac{\text{Rise}}{\text{Span}}$$
$$= \frac{12}{24} = \frac{1}{2}$$

Fig. 2-18. Pitch of a Roof

In structural engineering, the slope of a beam or other structural member is expressed as shown in Fig. 2-17 above.

The pitch of a roof in architectural drafting would be specified as shown in Fig. 2-18 above.

2.3.1-C To DRAW the VIEWS of a LINE GIVEN the TRUE LENGTH, SLOPE and BEARING

Analysis: It should be remembered that bearing can only be shown on the plan view and that the direction of north is always assumed to be toward the top of the drawing sheet. Since the bearing of the line is given, the line, indefinite in length, may be drawn in the plan view completely independent of the slope and true length. An auxiliary elevation view having lines of sight perpendicular to the bearing will show the true length and slope of the line. Once the true length is established in the auxiliary elevation view, the line may be projected back to the plan view. The front view can now be projected from the plan view with the distances taken from the auxiliary elevation view.

Example: Line *AB* bears N 45° E, is 420' long and has a downward grade of 30% from *A* to *B*. Construct the plan and front elevation views of the line.

Fig. 2-19 below shows the four steps involved in the solution of the problem.

Step 1: Establish point *A* in both the plan and front views. From point *A* in the plan view, draw a line having a bearing of N 45° E. This line is of an indefinite length. Construct a folding line *H*-1 parallel to the bearing line.

Step 2: Locate point *A* in the auxiliary elevation view, placing it the same distance away from the folding line as it is in the front view. From point *A* in view 1, draw a light line parallel to *H*-1 and measure off 100 units using any convenient scale. Measure a perpendicular distance of 30 units to locate a point on a line established as 30% downward grade. This line is also indefinite in length. (If the slope angle is given instead of the per cent grade, then a protractor should be used to lay out the angle.)

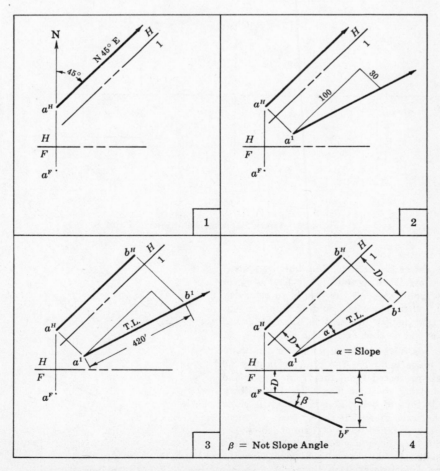

Fig. 2-19. Locating a Line of Given Bearing, Grade, and True Length

Step 3: From A along this 30% grade line of indefinite length, measure 420' to locate point B. This view now gives us both the true length and grade of the line AB. Project point B back to the bearing line on the plan view.

Step 4: To locate point B in the front view, project from the plan view down to the front view and step off the same distances for point B in the front view as we have on the auxiliary elevation view 1. Similar points on all elevation views will be the same distance below the horizontal image plane.

Note: A common error made by many students is that of laying out the slope angle or grade in the front view. This can only be done when the bearing of the line is due east and west. Remember that in order to see the slope or grade of a line, the line must appear in its true length in an elevation view.

2.3.2 The STRAIGHT LINE APPEARING as a POINT

Analysis: A line will appear as a point in any view in which the lines of sight are parallel to the line in space. The point view of a line represents every point on the line. The line must first appear in its true length in the view adjacent to that which shows it as a point. Usually the line is represented in its true length before it is projected as a point, but occasionally the point view of the line is established first and then the true length and other relative projections are made.

Example: In Fig. 2-20(*a*) below a vertical line projects as a point in the plan view; at (*b*) a level line projects as a point in the front elevation view; at (*c*) line AB is a level line appearing as a point in the profile view.

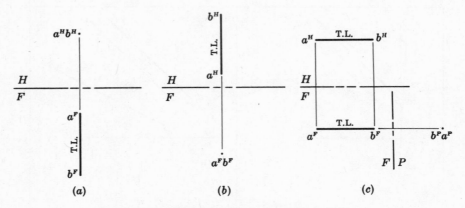

Fig. 2-20. Views Showing a Line as a Point

Fig. 2-21 shows an oblique line which requires four views to show it as a point. The first step in determining the point view of the line is to establish its true length. Then place the folding line perpendicular to the true length of the line. The resulting projection of the line is that of a point.

The true length view can either be an auxiliary elevation view projected from the plan or it may be an inclined view projected from the front view. The advantage of showing the true length in an auxiliary elevation view is that of being able to show the slope of the line in the same view if it is required. As mentioned before, the inclined view 3 cannot show the slope of the line because a horizontal plane does not appear as an edge in this view and slope *must* be measured from a horizontal base line.

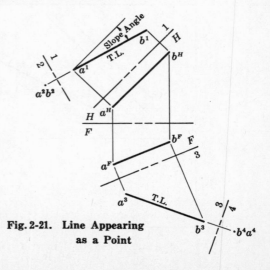

Fig. 2-21. Line Appearing as a Point

2.3.3-A The PLANE APPEARING as an EDGE

Definition: A plane is a surface such that a straight line connecting any two points in the surface lies entirely on that surface. Therefore any two lines in a plane will either be parallel or will intersect. For the purpose of problem solving, a plane can be considered as being of an indefinite extent.

Planes not perpendicular to any of the three principal planes of projection are called oblique planes.

In general engineering practice, there are four ways of representing a plane in multi-view drawing. The problems in this text can be solved by using any or all of these plane representations. In Fig. 2-22 below the four representations of planes are shown as follows: (*a*) Two intersecting lines*, (*b*) Two parallel lines, (*c*) Three points not in a straight line, and (*d*) A point and a line.

Note: In order for a plane to be represented as intersecting lines, the intersection point must project between related views as a common point.

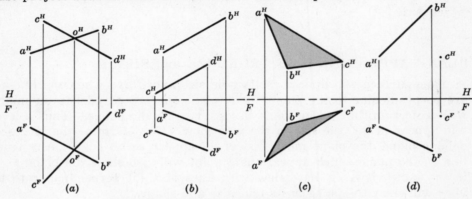

Fig. 2-22. Representations of a Plane Surface

Analysis: (1) The edge view of a plane should never be visualized as being a line but rather as a plane in which points are located at varying distances from the observer. (2) The method of showing an oblique plane as an edge is that of drawing an auxiliary view with lines of sight parallel to a horizontal, frontal, or profile line on the plane. (3) The line on the plane must first appear in its true length. (4) Any plane will appear as an edge, or straight line, in that view which shows any line in that plane as a point.

Example: To obtain the edge view of plane *ABC* in Fig. 2-23, draw a level line *AD* in the front view. Project to the plan view where the level line will now appear in its true length. Locate the folding line *H*-1 perpendicular to the true length line. Project the points from the plan view to the auxiliary elevation view. The true length line in the plan view now appears as a point. Project also points *B* and *C* in the auxiliary view. The straight line joining these two points (and the point view of line *AD*) determines the edge view of plane *ABC*.

If an innumerable number of level lines were drawn in the front view and shown as true length lines in the plan view, they would all project as points in the auxiliary elevation view. So actually, the edge view of a plane is an innumerable number of points which represent an innumerable number of true length lines.

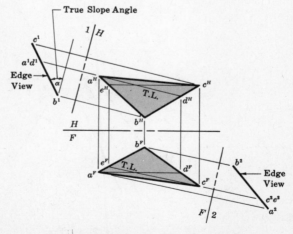

Fig. 2-23. Edge View of a Plane

An edge view of the plane can also be shown in an inclined view by drawing a frontal line *CE* in the plan view, showing it in its true length in the front view and then as a point in the inclined view which is projected from the front view.

2.3.3-*B* The TRUE SLOPE of a PLANE

Analysis: The slope angle of any plane is the angle that the plane makes with a horizontal reference plane. This slope angle may be measured in degrees or in per cent grade, just as the slope of a line is measured.

Example: In Fig. 2-23, the true slope angle of the plane can be seen only in the elevation view which shows the plane as an edge. The angle between this edge view and a horizontal reference plane is the true slope.

Note: Even though an inclined view may show the edge view of a plane, it cannot show the true slope because it cannot show a horizontal reference plane as an edge.

In geology and mining engineering, the term "dip" is used to express the slope angle of a plane. (See Art. 9-1)

2.3.4 The PLANE APPEARING in its TRUE SIZE and SHAPE

Analysis: A plane surface will appear in its true size and shape when the lines of sight are perpendicular to the plane. Any plane surface parallel to an image plane will project on that image plane in its true size and shape. Thus a level plane will appear in its true size in the plan view. To determine the true size of an oblique plane, the plane must first of all appear as an edge. Any true length line on the plane which appears as a point will also show the plane as an edge in the same view. A new view with lines of sight perpendicular to the edge view will show the plane in its true size and shape.

Example: To obtain the true size of plane *ABC* as shown in Fig. 2-24 below, the first step would be to show a true length line on the plane. This can be done by either drawing a frontal line *CE* in the plan view or a level line *CD* in the front view. The auxiliary elevation view 1 will show the level line as a point and, thus, the plane appears as an edge. The inclined view 2 shows the true size of the plane.

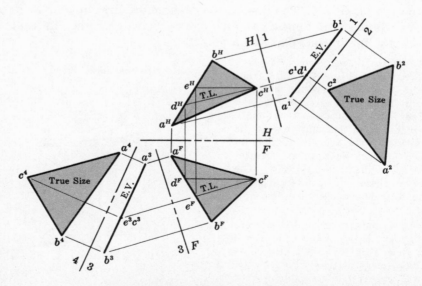

Fig. 2-24. True Size of an Oblique Plane

If a frontal line is used, its true length will appear in the front view, and view 3, an inclined view, will show the frontal line as a point and the plane as an edge. The inclined view 4 shows the plane in its true size and shape. View 4, of course, reveals the plane to be the same size and shape as it appears in view 2. This constitutes an excellent check for the accuracy of the work.

Quite often a problem requiring the finding of a plane in its true size can be simplified by using a line already on the plane in its true length. Fig 2-25 shows a plane, one line of which, *AB*, already appears in its true length. In this case, the line *AB* is shown as a point in view 1 which also reveals the plane as an edge. Subsequent projection in view 2 determines the true size of the plane.

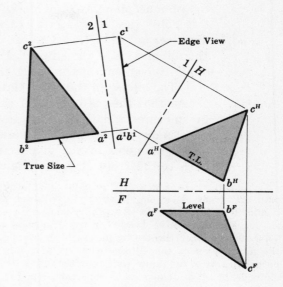

Fig. 2-25. Plane with Given True Length Line

2.4 To LOCATE a GIVEN PLANE FIGURE in a GIVEN PLANE

Analysis: A view showing the given plane in its true size will also show the true size and shape of the given plane figure. As mentioned previously, the true size view of a plane must be adjacent to a view showing the plane as an edge and the line of sight must be perpendicular to the edge view.

Example: In Fig. 2-26 below the plane *ABCD* and point *X*, the center of a 1″ square, are given in both the plan and front elevation views. Two sides of the square are to be parallel to the longest sides of the plane which is represented by parallel lines. Draw an auxiliary elevation view showing the plane as an edge. From this edge view, draw an inclined view which will show the true size of the plane *ABCD* and will also locate point *X* on the plane. Using *X* as the center, draw the required 1″ square with two sides parallel to *AB* and *CD*. The four corners of the square can now be projected back to the edge view and thence to the plan and front elevation views.

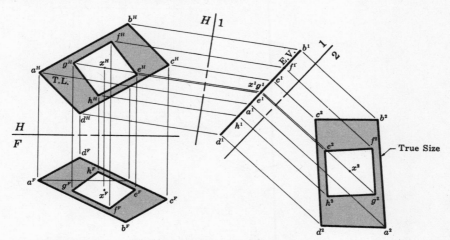

Fig. 2-26. To Locate a Given Plane Figure on a Given Plane

2.5 To DRAW a CIRCLE of GIVEN DIAMETER on an OBLIQUE PLANE

Analysis: A circle will appear as a circle or as an ellipse in every view except the view showing the circle as an edge. A level diameter will appear true length in the plan view and therefore will be the major axis of the ellipse. A view showing the oblique plane as an edge will also show the major axis as a point. Using this point view of the axis as the center of the circle, measure the actual given diameter of the circle. Project this diameter back to the plan view to determine the minor axis of the ellipse. Using the trammel method (see Art. A.5), or any other convenient method, the ellipse can now be completed in the plan view.

The ellipse in the front elevation view will be obtained by the same method except that the major axis of the ellipse will now be a frontal line through the given center and equal in length to the diameter of the given circle. An inclined view is drawn showing the plane as an edge, the major axis as a point, and the actual diameter of the given circle is again measured. Project this diameter back to the front elevation view to obtain the minor axis of the ellipse.

Example: In Fig. 2-27, the plane *ABC* and the center of a circle, point *X*, are given in both the plan and front elevation views. It is required to locate a circle of given diameter in the plane *ABC*. Again, it is necessary to show the plane as an edge, but not necessary to show it in its true size and shape. Having point *X* located in the edge view of the plane, measure the actual diameter of the circle. This diameter, represented by line *1-2*, can now be projected back to the plan view where it will determine the minor axis of the ellipse in this view. The minor axis will be located through the point *X* and perpendicular to the major axis represented by line *3-4*. Using the trammel method, or any other standard method, the ellipse can now be constructed in the plan view.

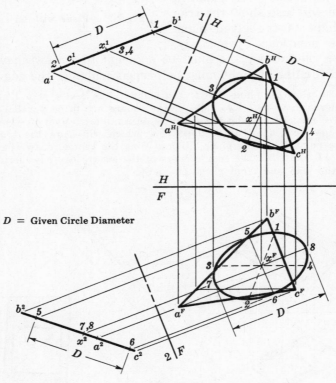

D = Given Circle Diameter

Fig. 2-27.

To complete the front elevation view, draw an inclined view 2 showing the plane as an edge. Locate point *X* in this view and lay out the true diameter of the given circle. Through point *X* in the front view, lay out the major axis of the ellipse equal to the actual diameter of the given circle. The minor axis, represented by line *5-6*, can now be projected from view 2 back to the front elevation view where it will be located perpendicular to the major axis *7-8* and through point *X*. Having both the major and minor axes determined, the ellipse can now be constructed as in the plan view.

Note: It should be noted that the major and minor axes in one view do not necessarily correspond to the major and minor axes in related views. However a test for accuracy is that of projecting the extreme left and right points on the ellipse from the plan view to the front elevation view.

Solved Problems

1. **Given:** Plan and profile views of the line AB in Fig. 2-28.   Scale: $\frac{1}{4}'' = 1'-0''$.

    Problem: Determine the true length and slope of the line AB.

    Solution:

    From both the plan and profile views project line AB into the front view. Draw an auxiliary elevation view with the folding line H-1 parallel to the plan view of the line. View 1 will show both the true length and slope of the line AB.

    Ans. T.L. $= 5'-11''$, Slope $= 39°$

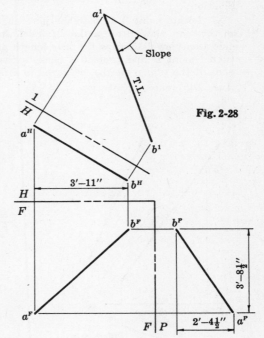

Fig. 2-28

2. **Given:** AB is the centerline of a pipe which runs parallel to a vertical building wall. See Fig. 2-29 below. The true distance between A and B is 150', and point B is 75' lower than A. A pipe connection is located 50' from B.   Scale: $1'' = 100'$.

    Problem: Locate the point of intersection in all three principal planes of projection. What is the elevation of the pipe connection if the elevation of point A is 350' above sea level?

    Solution:

    Using the given data, lay out AB as a frontal line having its true length shown in the front elevation view. In the front view, measure along the line from B a distance of 50'. Label this point C. Draw a profile view of line AB and project point C into this view. Project point C also into the plan view. The elevation of point C can be measured in either the front or profile view.   *Ans.* Elevation $= 300'$ above sea level

3. **Given:** Plane ABC in Fig. 2-30 below.   Point B is 4' east, 2' south of A and 6' below A. Point C is 7' east, 3' north of A and 2' above A. Point X on the plane is 2' east, 1' south of A and 3' below A.   Scale: $\frac{1}{8}'' = 1'-0''$.

    Problem: What is the bearing of a level line passing through point X?

    Solution:

    Draw the plan and front elevation views of plane ABC and point X. Draw a level line through point X in the front view until it intersects line BC. Label the intersection point Y and project it up to the plan view. Measure the bearing of line XY in the plan view.

    Ans. Bearing $XY = $ N 73° E

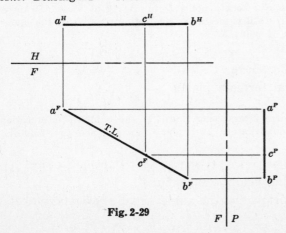

Fig. 2-29

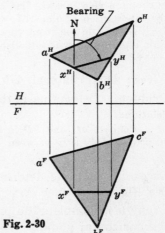

Fig. 2-30

4. **Given:** A tunnel at the bottom of a vertical shaft bears N 35°15′ W. The tunnel has a downward grade of 20% and is 128′ long. Scale: $1'' = 80'$.

Problem: How would the tunnel appear in the plan and front elevation views? What is the difference in elevation between the two ends of the tunnel?

Solution:

Locate point A in both the plan and front elevation views. See Fig. 2-31 below. From point A in the plan view, draw a line of indefinite length having a bearing of N 35°15′ W. Draw auxiliary elevation view 1 to show the true length and grade of the tunnel from A. Label the lower end of the tunnel point B and project it back to the plan and front views. The vertical distance between the ends of the tunnel can be measured in either the front view or auxiliary elevation view 1.

Ans. Difference in Elevation = 25′

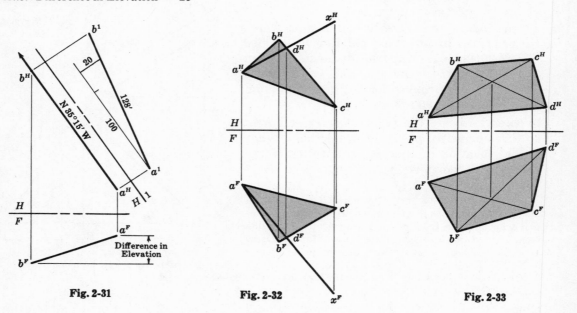

Fig. 2-31   Fig. 2-32   Fig. 2-33

5. **Given:** Plane ABC in Fig. 2-32 above. Point B is 3′ east, 3′ north of A and 5′ below A. Point C is 8′ east, 3′ south of A and 2′ below A. Point X in plane ABC has a map distance of 9′ away from A and bearing N 60° E from A. Scale: $\frac{1}{8}'' = 1'-0''$.

Problem: Locate point X in the front elevation view.

Solution:

Using the given data, draw the plan and front elevation views of plane ABC. From point A in the plan view, draw a line having a bearing of N 60° E and having a plan view length of 9′. Label the intersection of line AX with BC as point D. Locate point D in the front view. Draw a line from point A in the front view through point D until it meets the projection of X from the plan view. This intersection will establish the front view of point X.

6. **Given:** Quadrilateral $ABCD$ in Fig. 2-33 above.

Problem: Determine if the quadrilateral forms a plane.

Solution:

Connect AC and BD in both views to form intersecting lines. If the point of intersection can be projected directly between views then the quadrilateral is a plane figure. (See Art. 2.3.3-A)

7. **Given:** A plane is determined by $ABCD$ [$A(1,1,4)$ $B(1\frac{1}{2}, 2\frac{1}{4}, 3)$ $C(4, 1\frac{1}{2}, X)$ $D(3\frac{1}{2}, \frac{1}{2}, 4\frac{3}{4})$]. See Fig. 2-34. (See Art. 1.7 for coordinate problem layout.)

Problem: Complete the plan view and determine the angle which the plane makes with the horizontal plane.

Solution:

To locate point C in the plan view represent the plane by means of intersecting lines. Draw an auxiliary elevation view showing the plane as an edge. This elevation view 1 will show the slope of the plane.   *Ans.* Slope = 50°

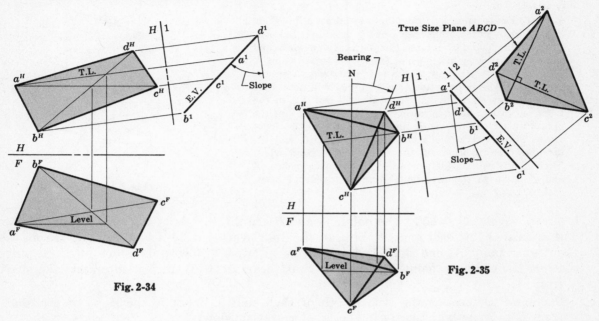

Fig. 2-34

Fig. 2-35

8. **Given:** The diagonals of a plane are represented by the lines AB and CD [$A(1, 2\frac{1}{2}, 5\frac{1}{2})$ $B(3, 2, 5)$ $C(2, 1\frac{1}{4}, 3\frac{3}{4})$]. The diagonals are of equal length and intersect each other at right angles.   Scale: $\frac{1}{8}'' = 1'-0''$.

   Problem: Determine the true size and slope of plane $ABCD$. Find the true length and bearing of the line CD. Complete the plan and front views of the plane.

   Solution:

   Connect points A, B, and C in both the plan and front elevation views as shown in Fig. 2-35 above. Draw an auxiliary elevation view showing the plane as an edge. The slope of the plane is measured in this view. Draw an inclined view showing the plane ABC in its true size. From point C draw a perpendicular to AB. Make this line CD equal in length to the line AB in view 2. The plane $ABCD$ appears true size in view 2. Project point D back to the other views and connect it to point C. The bearing of CD will be measured in the plan view. (See Art. 2.3.4)

   Ans. T.L. = 8'–6", Slope of plane = 33°30', Bearing = N 22° E

9. **Given:** A pipeline is 160′ long, bears S 45° E, and slopes downward 20°. The highest end of the line is at elevation 200′.   Scale: $1'' = 100'$.

   Problem: Draw the plan and front elevation views of the pipeline. What is the elevation at the lower end?

   Solution:

   From point A in the plan view, draw a line of indefinite length and having a bearing of S 45° E. See Fig. 2-36. Place folding line H-1 parallel to the plan view of the line from A. Project into auxiliary elevation view 1 where line AB will appear in its true length of 160′ and will have a slope of 20°. Project point B to the plan view. Points A and B can now both be located in the front elevation view. The elevation of point B can be measured in either the front elevation view or auxiliary elevation view 1.   *Ans.* Elevation B = 145′

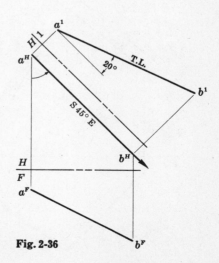

Fig. 2-36

10. Given: Plane ABC in both views, line YZ in the plan view and point X in the front view. Line YZ and point X lie on the plane of ABC. See Fig. 2-37.

Problem: Complete the plan and front elevation views.

Solution:

To locate point X in the plan view, draw a line from A through X in the front view until it intersects line BC. Project this point of intersection to the plan view and show the plan view of the line from A. Project point X from the front view up to the new line from A in the plan view. The intersection thus incurred will determine the plan view of point X. To locate the line YZ in the front elevation view, project the intersections in the plan view of the line with the plane down to the front view. These intersections on lines AC and BC will determine the direction of the line in the front view. Their limits will be determined by projecting points Y and Z down to the directional line in the front view. (See Art. 2.2)

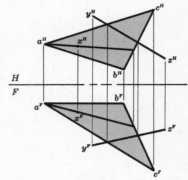

Fig. 2-37

11. Given: Scale: $1'' = 80'$. The points A and B are the portals of two coal mining shafts. A is located 120' east and 40' south of B. Its elevation is 30' lower than B. From the entrance point A, one shaft bears N 30° W and has a downward slope of 20 degrees. From the entrance point B, the other shaft bears N 50° E until it intersects the shaft from A.

Problem: Determine the true length of each shaft. What is the slope of the shaft from B? Show both shafts in the front elevation view.

Solution:

Using the given data, locate points A and B in both the plan and front elevation views as shown in Fig. 2-38 below. Locate the intersection of the bearing lines in the plan view. Label this intersection point C. Draw an auxiliary elevation view 1 to determine the true length of the shaft from A to the intersection C. Draw an auxiliary elevation view 2 to obtain the true length and slope of shaft BC. Project point C into the front view and locate its elevation as established by view 1.

Ans. T.L. of $AC = 116'$, T.L. of $BC = 110'$, Slope of $BC = 39°$

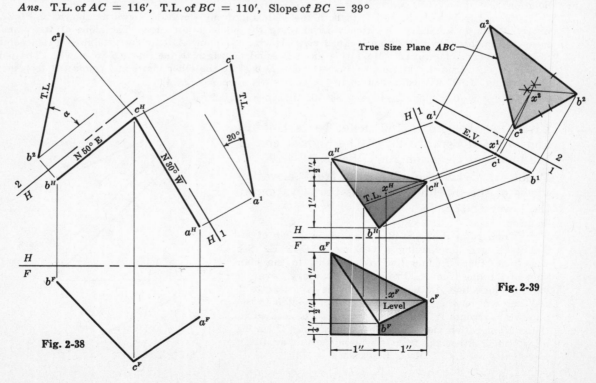

Fig. 2-38

Fig. 2-39

12. Given: Fig. 2-39 above shows the plan and front elevation views of a truncated triangular prism.   Scale: $\frac{1}{2}'' = 1''$.

Problem: Determine the location of a point X on the oblique plane ABC which will be equidistant from the three edges of the plane.   How far above the base of the prism is point X?   Show point X in all views.

Solution:

Draw an auxiliary elevation view showing plane ABC as an edge.   From this view project an inclined view to show the true size of the plane.   Using angle bisectors, locate the point X equidistant from all three edges of the plane.   Project point X back to the other views.   In the front view measure the vertical distance from point X to the base of the prism.   *Ans.* $\frac{13}{16}''$

13. Given: Fig. 2-40 shows the structural framework between two adjacent buildings.   Scale: $\frac{1}{16}'' = 1'-0''$.

Problem: Determine the true length of structural members AB and CD.   What is the true length and per cent grade of member BC?

Solution:

Place a folding line H-1 parallel to the plan view of $CABD$.   Project to obtain an auxiliary elevation view.   This auxiliary elevation view will yield the true length and grade as required.

Ans. T.L.: $AB = 11'-3''$, $CD = 13'-3''$, $BC = 7'-2\frac{1}{2}''$.
  Grade $BC = 18\%$

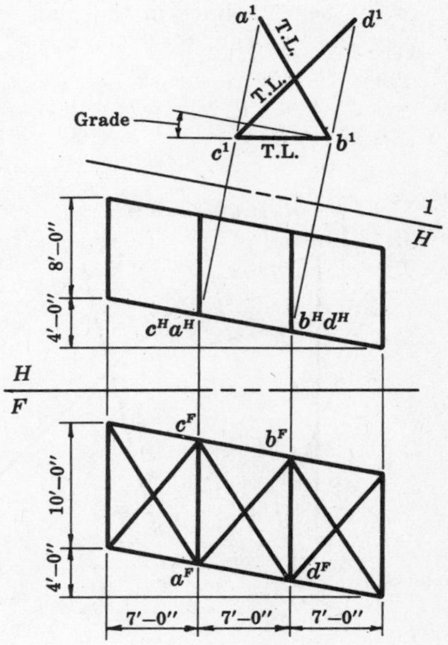

14. Given: A hip roof which must be covered with asbestos shingles is shown in Fig. 2-41 below. Scale: $1'' = 30'$.

Problem: Show the true size of planes A and B. How many "squares" of shingles will be required to cover the roof?   (1 Square $= 100$ Sq. Ft.)

Solution:

Since plane B is shown as an edge in the front elevation view, a folding line F-1 is placed parallel to this edge view in order to obtain an inclined view which will show the true size of plane B.   Place a folding line F-P or H-P perpendicular to folding line H-F in order to obtain an edge view of plane A.   Place a folding line parallel to the edge view of plane A and project to obtain the true size of plane A.   Calculate the area of planes A and B, then multiply by two since there are two of each size planes.   *Ans.* 12 Squares

Fig. 2-40

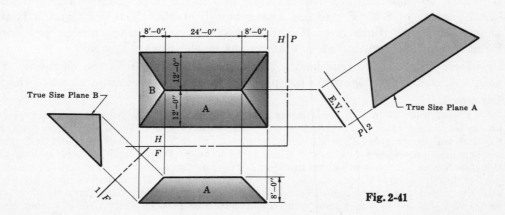

Fig. 2-41

15. Given: *AB* and *CD* are the centerlines of two conveyor ducts having circular cross sections. One duct bears N 45° W from *A*, is 75′ long, and has a slope from *A* of −20°. Point *C*, which is located 50′ due west of *A* and at the same elevation as *A*, is the beginning of a duct bearing N 15° E, which is 55′ long and has a slope of −15° from *C*. Both ducts have a diameter of 3′−0″. See Fig. 2-42 below. Scale: 1″ = 40′.

Problem: What is the vertical clearance between the two ducts?

Solution:

Locate the points *A* and *C* in the plan view. Draw lines of indefinite length from both *A* and *C* having the given bearings. Draw auxiliary elevation view 1 to determine the end of the duct which begins at *A*. Label this point *B* and project it to the plan view. Draw auxiliary elevation view 2 in order to show the true length and slope of the duct from *C*. Label the other end of the duct, point *D*. Project points *A* and *B* from the plan view into auxiliary elevation view 2. The intersection of the bearing lines in the plan view will determine the vertical distance, or clearance, in elevation view 2.   *Ans.* 6′−6″

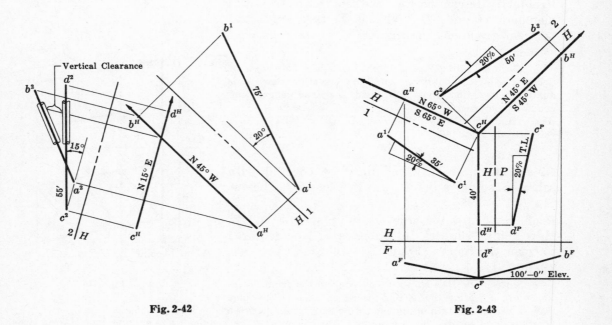

Fig. 2-42   Fig. 2-43

16. Given: Three terra cotta pipelines converge at a manhole *C* from *A*, *B*, and *D*. Point *D* is located 40′ due south of the manhole and has a 20% grade. The pipeline from *A* has a bearing of S 65° E and has a true length of 35′. The line from *B* bears S 45° W and is 50′ long. Both lines *A* and *B* have the same grade as the pipeline from *D*. Point *C* is at an elevation of 100′ above sea level. Scale: 1″ = 40′.

Problem: What is the true length of pipeline *CD*? What is the elevation of the three points *A*, *B*, and *D*? Show the complete plan and front elevation views of the sewer line.

Solution:

Locate line *CD* in the plan view as well as the bearing lines from *C* on which points *A* and *B* are to be located. See Fig. 2-43 above. A profile view of line *CD* will determine the true length of *CD* and will locate point *C* in all elevation views. Place folding lines *H*-1 and *H*-2 parallel to the bearing lines on which points *A* and *B* are to be located. Project point *C* into each of the auxiliary elevation views. Lay out the given grade and true lengths of lines *CA* and *CB*. From views 1 and 2, project points *A* and *B* back to the plan and front views. The elevation of points *A*, *B*, and *D* relative to point *C* can be measured individually in the elevation views showing their true lengths, or they may all be measured from a horizontal base line through point *C* in the front elevation view.
Ans. T.L. *CD* = 41′−0″. Elevation *A* = 106′−6″, *B* = 110′−0″, *D* = 108′−6″

17. Given: Two mine tunnels start at a common point A in a vertical shaft as shown in Fig. 2-44 below. Tunnel AB is 160′ long, bears S 42°15′ E on a downward slope of 24°. Tunnel AC is 110′ long, bears N 40°30′ E on a downward slope of 18°. Scale: 1″ = 100′.

Problem: If a new connecting tunnel between points B and C were dug, what would be its length, bearing, and per cent grade?

Solution:

Using the given data, locate point A in the plan view and draw lines of indefinite length having the given bearings. Draw auxiliary elevation view 1 showing the tunnel from A in its true length. Project the end B of this tunnel back to the plan view. Draw auxiliary elevation view 2 showing the second tunnel from A in its true length. Project the end C of this tunnel back to the plan view also. Connect points B and C in the plan view and measure the bearing of the proposed connecting tunnel. Draw an auxiliary elevation view 3 which will show both the true length and grade of the proposed connecting tunnel from B to C.

Ans. T.L. BC = 192′,   Bearing BC = N 9° W,   Per Cent Grade BC = 15%

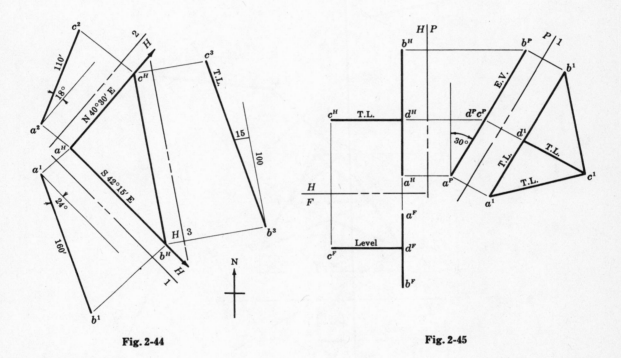

Fig. 2-44   Fig. 2-45

18. Given: A mine shaft AB slopes downward 30° from A. Point B is located 55′ due north of A. It is proposed to connect the mine shaft with a level ventilating duct from point C in another shaft. Point C is 30′ west and 25′ north of A. Points A, B, and C lie in a plane perpendicular to the profile plane. See Fig. 2-45 above. Scale: 1″ = 40′.

Problem: Locate point C in the front elevation view. What will be the true length of the ventilating duct? At what distance from point A will the duct enter the shaft AB? What is the true distance from C to point A?

Solution:

Using the given data, locate the plan view of the shaft and point C. Draw a profile view showing the slope of plane ABC. Line AB and point C can now be located in the front elevation view. An inclined view 1 will show the true size of the plane ABC. All measurements can be made in this view even though the true length of the ventilating duct can be obtained in the plan and front views.

Ans. T.L. CD = 30′–0″,   Distance from A = 29′–0″,   Distance from C to A = 41′–6″

19. Given: Fig. 2-46 below shows a symmetrical "A" frame which is being used in an industrial plant as a structural support. Scale: $\frac{3}{16}'' = 1'-0''$.

Problem: What is the true length and slope of structural members AB, AF, and BF?

Solution:

Draw folding lines H-1, H-2, and H-3 parallel to the plan view of lines AB, AF, and BF, respectively. Project into these auxiliary elevation views to obtain the true length and slope of lines AB, AF, and BF.

Ans. T.L. $AB = 11'-10''$, Slope $AB = 57°30'$
T.L. $AF = 8'-6''$, Slope $AF = 36°$
T.L. $BF = 6'-6''$, Slope $BF = 50°30'$

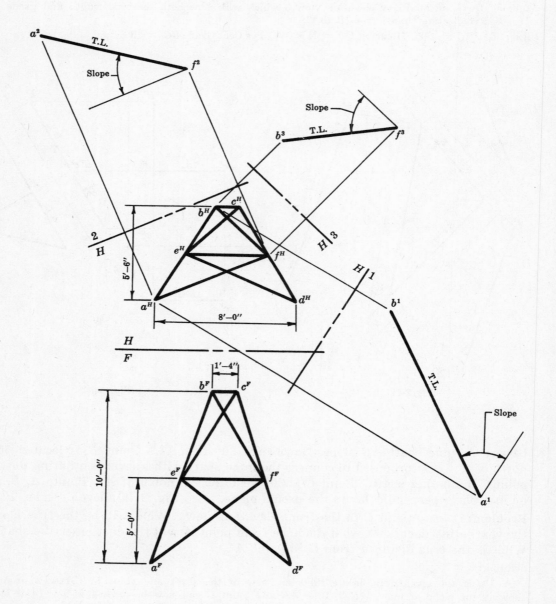

Fig. 2-46

20. Given: A 12′ high television antenna *AB* is erected atop an apartment building. Its stability is maintained by three guy wires located as follows: *C* is 8′ south, 7′ east of *AB* and is 2′ higher than *B*. *D* is 4′ south, 9′ west of *AB* and is fastened to an anchor 3′ above *B*. Point *E* is 8′ north, 2′ east and at the same elevation as *B*. See Fig. 2-47 below. Scale: $\frac{1}{8}'' = 1'-0''$.

Problem: What is the true length and bearing of each guy wire? What angle does each of the wires make with the vertical antenna? Show the antenna in all views.

Solution:

Using the given data, show the antenna and guy wires in both the plan and front elevation views. Measure the bearing of each guy wire in the plan view. Draw auxiliary elevation views 1, 2, and 3 to show the true lengths of the guy wires and antenna. Measure the angle formed by the guy wire and the antenna in each of the auxiliary elevation views.

Ans. Guy Wire *AC*: T.L. = 14′–6″, Bearing = S 41° E, Angle *α* = 47°
　　　Guy Wire *AD*: T.L. = 13′–4″, Bearing = S 66° W, Angle *β* = 48°
　　　Guy Wire *AE*: T.L. = 14′–6″, Bearing = N 15° E, Angle *γ* = 34°

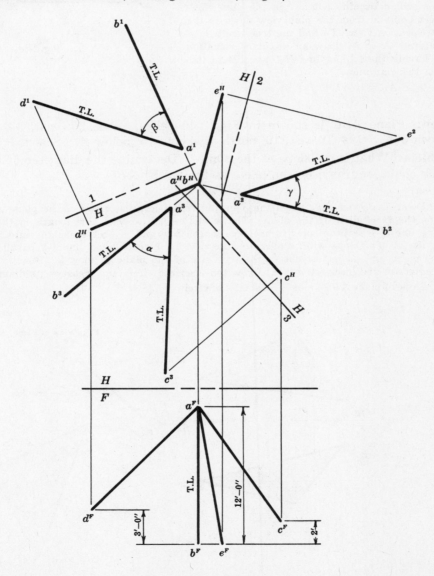

Fig. 2-47

21. Given: Plane *ABC* is shown in Fig. 2-48. *B* is 25′ west, 20′ south of *A* and at the same elevation as *A*. *C* is 12′ west, 20′ south of *A* and 15′ above *A*. Scale: 1″ = 20′.

Problem: Using two views only, locate point *X* on the plane at an elevation of 5′ above *A* and 10′ south of *A*. Show this point *X* in both views. Determine the true distance from *A* to point *X*.

Solution:

Draw the plan and front elevation views of plane *ABC*. At a distance of 5′ above *A* in the front view, draw a level line on the plane. Show the plan view of this level line intersecting a frontal line drawn 10′ south of *A*. The intersection will determine point *X* in the plan view. Project point *X* from the plan view down to the front elevation view. To find the true length distance from *A* to *X*, draw an auxiliary elevation view having the folding line *H*-1 parallel to line *AX* in the plan view.

Ans. T.L. *AX* = 13′–10″

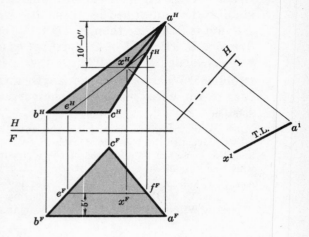

Fig. 2-48

22. Given: Plane *ABC* is shown in Fig. 2-49 below. *B* is 20′ east, 30′ north of *A* and 25′ below *A*. *C* is 60′ east, 10′ north of *A* and 15′ below *A*. Scale: 1″ = 40′.

Problem: What is the slope of the plane? Determine the diameter of the largest circle which could be drawn on the plane as limited by *ABC*.

Solution:

Using the given data, draw the plan and front elevation views of the plane *ABC*. Draw a level line on the front view of the plane. The level line will appear true length in the plan view. Place folding line *H*-1 perpendicular to the true length line and project the plane into the auxiliary elevation view 1. The slope will be measured in view 1. Place folding line 1-2 parallel to the edge view of the plane and project to obtain the true size of the plane in view 2. Bisect the angles to locate the center for the inscribed circle. Draw the inscribed circle and measure its diameter in view 2.

Ans. Slope of plane *ABC* = 38°, Diameter of circle = 26′–0″

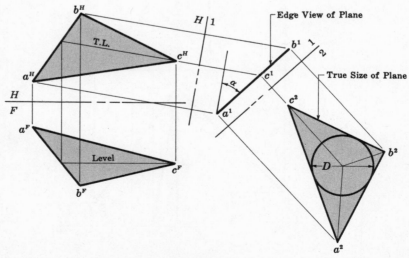

Fig. 2-49

23. Given: Two "feeder line" sanitary sewers start from a common point *A* in a manhole. See Fig. 2-50 below. Sewer line *AB* is 50′ long, falls 20% and has a bearing of N 60° W. Sewer line *AC* is 45′ long, falls 15° and bears due south. Both of these sewer lines terminate at the main "trunk" sewer line which leads to a river. Scale: $1'' = 40'$.

Problem: How far are points *B* and *C* from each other? What is the bearing of the main sewer line between points *B* and *C*? Determine the slope of the sewer line from *B* to *C* in degrees.

Solution:
    Using the given data, from point *A* in the plan view, draw lines of indefinite length in the direction of the given bearings. Draw a profile view showing the true length and slope of the sewer line *AC*. Project point *C* to the plan view. Draw an auxiliary elevation view showing the true length and slope of the sewer line *AB*. Project point *B* to the plan view. Connect points *B* and *C* in the plan view and measure the bearing. Place folding line *H-2* parallel to the plan view of *BC* and project into the auxiliary elevation view 2 in order to obtain the true length and slope of *BC*.

Ans. True Length *BC* = 79′, Bearing *BC* = S 32° E, Slope *BC* = −4°

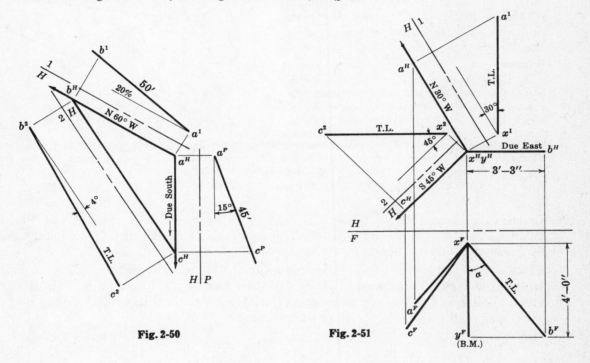

Fig. 2-50   Fig. 2-51

24. Given: The three equal legs of a surveyor's tripod are located in their relationship to the plumb line as follows: Leg *A* bears N 30° W and has a slope of 30 degrees; Leg *B* is extended 3′–3″ due east of the plumb line and at the same elevation as the bench mark. The plumb bob touches the bench mark at a vertical distance of 4′ below the top of the line. Leg *C* bears S 45° W and has a slope of 45°. Scale: $\frac{1}{4}'' = 1'-0''$.

Problem: Determine the true length of legs *A*, *B*, and *C*. What angle does leg *B* make with the plumb line? Show the three legs in both the plan and front views.

Solution:
    Draw partial plan and front elevation views. See Fig. 2-51 above. Since leg *XB* is a frontal line it will appear in its true length in the front view and thus will determine the true length of the other two legs, *XA* and *XC*. The angle formed by *XB* and the plumb line is measured in the front view. Pass folding lines *H-1* and *H-2* parallel to legs *XA* and *XC* respectively. In each of the two auxiliary elevation views lay out the given slope angle and measure the true length of each leg. Project points *A* and *C* back to the plan and front views.

Ans. T.L. of legs *A*, *B*, and *C* = 5′–2″; Angle between leg *B* and plumb line = 39°

25. Given: A production line conveyor system is presently laid out as indicated in the following chart:

Section	Bearing	Slope	Length
AB	N 45° W	−20%	120′
BC	Due West	0	90′
CD	S 25° W	−20°	135′

A revision in production methods calls for the elimination of the present conveyor system and the installation of a new conveyor directly from *A* to *D*. See Fig. 2-52. Scale: 1″ = 120′.

Problem: What would be the true length, slope in degrees, and bearing of the new conveyor? Draw the plan and necessary elevation views showing the conveyor system.

Solution:

Using the given data, and beginning with point *A* in the plan view, draw a bearing line from *A* at N 45° W of indefinite length. Draw an auxiliary elevation view 1 showing the true length and slope from *A* to *B*. Project point *B* back to the plan view. Repeat the procedure for lines *BC* and *CD* until point *D* is located in the plan view. Place folding line *H-3* parallel to the plan view of *AD* and project to obtain the true length and slope of line *AD*. The bearing of *AD* is measured in the plan view.

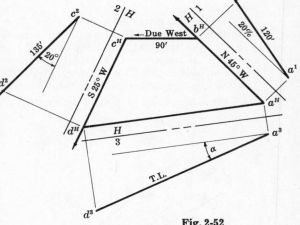

Fig. 2-52

Ans. T.L. *AD* = 239′, Slope *AD* = 17°, Bearing *AD* = S 82° W

26. Given: Two sewer lines *AB* and *CB* converge at a manhole *B* as shown in Fig. 2-53 below. Point *A* is 35′ north, 10′ east of *B* and 30′ above *B*. Point *C* is located 20′ north, 60′ west of *B* and 15′ above *B*. Scale: 1″ = 40′.

Problem: A new sewer line is to be located in the plane *ABC* and beginning at point *D* which is located 30′ due west of *A*. Using two views only, locate point *D* in the front view. Using as many views as necessary, determine the length of each sewer line that converges at *B*. What is the slope of the plane?

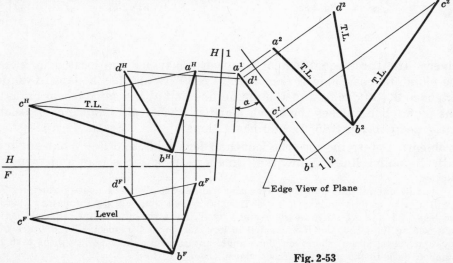

Fig. 2-53

Solution:

Locate plane ABC in both the plan and front elevation views. Locate point D in the plan view and draw a line to the plan view of point B. Project the intersection of line DB and AC in the plan view down to the front elevation view. From point B in the front view, draw a line through the projected point of intersection until it meets the projection of D from the plan. This will establish point D in the front view. Show the true length of a level line on the plane. Draw an auxiliary elevation view 1 showing the plane as an edge. The slope of the plane is measured in this view. Draw an inclined view 2 showing the plane in its true size. The true length of each sewer line is measured in this view.

Ans. T.L. $AB = 47'–0''$, T.L. $CB = 64'–6''$, T.L. $DB = 49'–0''$, Slope $= 40°$

27. **Given:** From a lighthouse 250′ above sea level, a submarine is sighted N 45° W of the observer and at an angle of depression of 20°30′. Five minutes later the submarine is spotted N 12° E and at an angle of depression of 15°15′. Scale: 1″ = 400′.

Problem: How far is the ship from the lighthouse at each sighting? How far did the submarine travel in five minutes? What was the course bearing and speed of the submarine if the course and speed were steady? (1 Knot = 6080′/hr.)

Solution:

Let point A represent the lighthouse as shown in Fig. 2-54 below. From A draw lines of indefinite length and having the given bearings. Place a folding line H-1 parallel to the bearing of the first sighting from the lighthouse. The auxiliary elevation view 1 will establish the location of the ship on the sea at first sighting. Project point B representing the ship back to the plan view. Locate folding line H-2 parallel to the bearing of the second sighting from the lighthouse. Auxiliary elevation view 2 will determine the distance from the lighthouse to the ship at second sighting. Project the location of the ship at second sighting back to the plan view. Connect the plan view of the ship locations at first and second sightings. This line will determine the course bearing of the submarine. Disregarding the curvature of the earth, this line will also show the distance the submarine travelled in five minutes. Calculate the speed by simple mathematics.

Ans. Distance—1st Sighting = 710′, Distance—2nd Sighting = 940′, Distance Submarine Travelled = 775′, Bearing = N 57°30′ E, Speed = Approx. 1.5 Knots

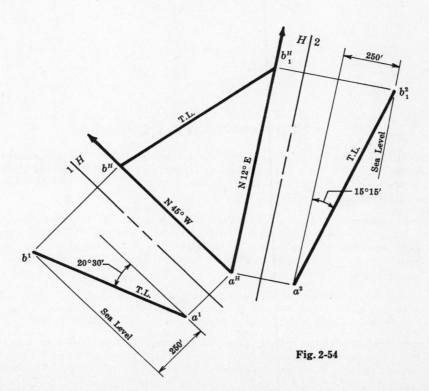

Fig. 2-54

28. Given: A rocket launcher bears due east, is 30′ long and has a slope of 30 degrees. The launcher is supported by two braces at its midpoint. The braces are perpendicular to the launcher, to each other, and are the same length as they rest on the level ground. See Fig. 2-55 below. Scale: 1″ = 20′.

Problem: What is the true length, slope, and bearing of these braces?

Solution:

Draw the complete front elevation view and a partial plan view. Draw an inclined view 1 showing the launcher braces in their true length. Locate points Y and Z in the plan view. Draw an auxiliary elevation view showing the true length of one of the braces. The slope can be measured in this view. Both braces will have the same slope. Their bearings are measured in the plan view.

Ans. Brace XY: T.L. = 12′–3½″, Slope = 38°, Bearing = S 26° E

Ans. Brace XZ: T.L. = 12′–3½″, Slope = 38°, Bearing = N 26° E

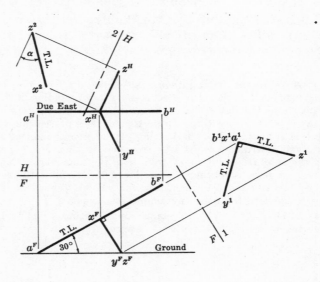

Fig. 2-55.

Supplementary Problems

29. A horizontal supporting bar is 12′ long and has a bearing of N 60° E. A vertical connection is made at the mid-point of the bar. Locate this connecting point in all three principal planes. Will the connection be shown at the mid-point of the bar in every view? Scale: ¼″ = 1′–0″.

Ans. Yes, except when the bar appears as a point.

30. The inclined railway to the top of Mt. Washington is 660′ long. The elevation at the top is 1250′ above sea level. The slope of the incline is 60°. If there are five "A" frame supports located equidistant along the incline, what would be the elevation of the connection for the middle support? Locate the five support connections in all three principal planes. Scale: 1″ = 100′.

Ans. Elevation = 964′

31. A stretch of straight highway is 155′ long, bears S 60°30′ W with a 10% downward grade. Show the centerline of this highway segment in the plan and front elevation views. Determine the difference in elevation between the two ends of the highway segment. Scale: 1″ = 50′.   *Ans.* 15′–6″

32. A surveyor's transit tripod is set up on a level highway. Leg AB is 6′ long and has a slope of 45°. Leg AC is 5′–3″ long. Leg AD has a slope of 60°. In the plan view one leg bears due south and the legs are equally spaced. Determine the slope of leg AC. How long is leg AD? Scale: ½″ = 1′–0″.

Ans. Slope AC = 54°, T.L. AD = 4′–11″

33. In each of the problems shown in Fig. 2-56 below the student will determine the true length and slope of the line *AB*. Measure to the closest sixteenth of an inch. Check the true lengths by mathematical calculations, using the method explained in Art. 2.3.1-*A*.

Ans. 1. T.L. $2\frac{3}{8}''$, Slope 13°   4. T.L. $2\frac{1}{8}''$, Slope 35°30′   7. T.L. 2′–2″, Slope 43°30′

2. T.L. $2\frac{5}{8}''$, Slope 35°   5. T.L. $3\frac{5}{16}''$, Slope 22°   8. T.L. $2\frac{1}{2}''$, Slope 22°

3. T.L. $1\frac{15}{16}''$, Slope 31°30′   6. T.L. $2\frac{1}{4}''$, Slope 26°30′   9. T.L. $2\frac{7}{16}''$, Slope 50°30′

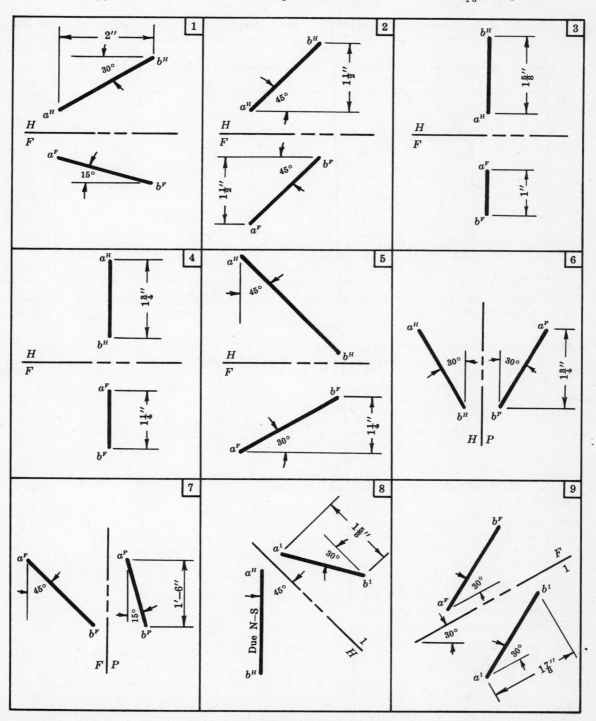

Fig. 2-56

34. An assembly line conveyor section begins on the second floor of a manufacturing plant. The start of the conveyor is located 24' west, 8' north and 16' above the other end of the conveyor. If the sloping distance from the elevation of the second floor to the false ceiling of the first floor is 11'–3'', and the conveyor starts 3' above the elevation of the second floor, what is the true length, slope, and bearing of the conveyor? How much of the conveyor will protrude above the second floor and below the first floor ceiling? Scale: $\frac{1}{8}'' = 1'-0''$

   Ans. T.L. = 30', Slope = 32°, Bearing = N 71°30' W, Above = 5'–8'', Below = 13'–0''.

35. In each of the following problems show the line in the plan and front elevation views. By graphical methods, determine the difference in elevation between the two ends of the line. Measure to the closest inch.

	Line	Bearing	Slope	True Length	Scale	Ans.
(1)	AB	N 30° W from A	Falls 20°	32'–6''	1'' = 10'	11'–1''
(2)	CD	N 15° E from C	Rising 50°	10'–6''	$\frac{1}{4}'' = 1'-0''$	8'–1''
(3)	EF	S 75° W from E	Rising 100%	6'–3''	$\frac{1}{2}'' = 1'-0''$	4'–5''
(4)	GH	Due east from G	Rising 0.5	130'	1'' = 40'	58'–0''
(5)	JK	S 35° E from J	Falling 45%	3'–8''	1'' = 1'-0''	1'–6''
(6)	LM	N 75° W from L	$+\frac{7}{10}$	65'–9''	1'' = 20'	37'–9''
(7)	NP	Due south from N	Rising 30%	12'–3''	$\frac{1}{4}'' = 1'-0''$	3'–6''
(8)	QR	N 45°30' E from Q	Falling 30°30'	27'	1'' = 10'	13'–8''

36. A plane is determined as follows: $A(1\frac{1}{2}, 2\frac{1}{4}, 6\frac{3}{4})$ $B(3, 4, 5)$ $C(4, 2\frac{1}{2}, 5\frac{3}{4})$. Scale: $\frac{1}{4}'' = 1'-0''$. Determine the diameter of a circle which would circumscribe the plane limited by ABC. What is the slope of the plane?   *Ans.* Diameter = 12', Slope = 56°

37. Two intersecting lines determine the plane $ABCD$. Point B is located 50' east, 70' north of A and 20' below A. Point C is located 10' west, 60' north of A and 30' below A. Point D is located 40' east of A and 30' north of A. Scale: 1'' = 20'. Show the true size of the plane as limited by $ABCD$. Determine the slope of the plane.   *Ans.* Slope = 27°30'

38. The inclined view 2 in Fig. 2-57 shows a 2'' diameter circle in its true size and shape. Scale: 12'' = 1'-0''. Project the circle back to the auxiliary elevation, plan, and front views. Determine the slope of the plane.   *Ans.* Slope = 30°

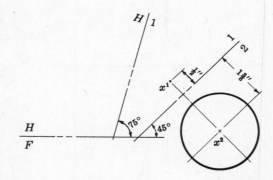

Fig. 2-57

In each of the following problems determine the edge view, slope, and true size of the plane:

39. Plane ABC. Point B is 25' east of A and at the same elevation as A. Point C is 20' east, 20' north of A and 10' above A. Scale: 1'' = 10'.
   Ans. Slope = 27°

40. Plane DEF. Point E is 20' east, 6' north of D and 3' above D. Point F is 2' west, 6' north of D and 4' below D. Scale: $\frac{1}{8}'' = 1'-0''$.
   Ans. Slope = 33°

41. Plane GHJ. Point G is 6' due south of H and 6' above H. Point J is 3' north, 4' west of G and 10' below G. Scale: $\frac{1}{4}'' = 1'-0''$.   *Ans.* Slope = 64°

42. Plane KLM. Point K is 20' due north of L and 35' above L. Point M is 40' due north of L and 10' above L. Scale: 1'' = 20'.   *Ans.* Slope = 90°

43. Plane NOP. Point N is 70' west, 45' north of O and 30' above O. Point P is 15' west, 20' north of O and 25' below O. Scale: 1'' = 40'.   *Ans.* Slope = 74°30'

44. Plane *QRS*. Point *R* is 15′ east, 20′ south of *Q* and 30′ above *Q*. Point *S* is 30′ east, 15′ north of *Q* and 10′ above *Q*. Scale: 1″ = 10′.   *Ans.* Slope = 50°

45. Plane *TUV*. Point *U* is 3′–6″ due south of *T* and 4′–3″ above *T*. Point *V* is 4′–9″ east of U, 2′–3″ north of *U*, and at the same elevation as *T*. Scale: ½″ = 1′–0″.   *Ans.* Slope = 51°30′

46. Plane *XYZ*. Point *Y* is 2′–9″ due east of *X* and at the same elevation as *X*. Point *Z* is 3′–6″ due east of *X* and 3′ below *X*. Scale: 1″ = 1′–0″.   *Ans.* Slope = 90°

In the following problems show the true size of plane *ABC*. Also determine the slope of the plane:
(See Art. 1.7 for coordinate problem layout)

47. $A(1, 1, 3)$ $B(1\frac{1}{2}, 2, 4\frac{1}{2})$ $C(3, 1, 3\frac{1}{2})$.   *Ans.* Slope = 37°

48. $A(1, 2, 3\frac{1}{2})$ $B(2, 1, 4\frac{1}{2})$ $C(3, 2\frac{1}{2}, 4)$.   *Ans.* Slope = 60°

49. $A(1, \frac{1}{2}, 4)$ $B(1, 2, 3)$ $C(2\frac{1}{2}, 1\frac{1}{2}, 4\frac{1}{2})$.   *Ans.* Slope = 62°

50. $A(7, 2\frac{1}{4}, 3)$ $B(6, 1, 4)$ $C(5, 2, 4\frac{1}{4})$.   *Ans.* Slope = 73°

51. $A(6\frac{1}{2}, 2, 4)$ $B(7, 1\frac{1}{2}, 5)$ $C(5\frac{1}{4}, 3, 4)$.   *Ans.* Slope = 45°

52. Determine the true size and slope of plane *ABCD*. Line *AD* is parallel to *BC*. Scale: ¼″ = 1′–0″. See Fig. 2-58 below.   *Ans.* Slope = 42°

53. The hip roof of a utility shed adjoining a barn is to be 12′ long, 8′ wide, and will begin at the barn and extend 8′–6″ along the ridge. The two corner rafters will have a slope angle of 45°. Scale: ¼″ = 1′–0″. What is the true length of the corner rafters? How much higher will the ridge of the roof be above the plane of the eaves? Show the plan and front elevation views of the roof.
Ans. T.L. = 7′–5½″, Higher by 5′–3″

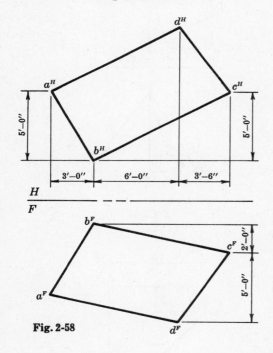

Fig. 2-58

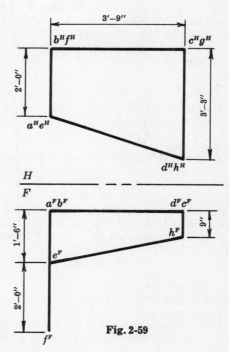

Fig. 2-59

54. Fig. 2-59 above shows the plan and partial front view of a metal transition piece. The bottom opening, *EFGH*, is a plane. Complete the front view. Scale: ½″ = 1′–0″. Determine the true size and slope of the plane *EFGH*.   *Ans.* Slope = 45°

55. Line *AB* is a vertical pole 8′ high. It is held in position by three guy wires which are located as follows: *AC* is 10′ long, has a 45° slope and a bearing of N 60° W. Guy wire *AD* is 8′ long, makes an angle of 30° with the pole and has a bearing of N 20° E. The wire from *E* is fastened to the pole 2′ from the top, is due south and makes an angle of 30° with the pole. If point *E* is at an elevation of 6″ above *B*, what is the true length of the wire from *E*? Scale: ¼″ = 1′–0″. Locate the three guy wires in the front and plan views.   *Ans.* T.L. = 6′–3½″

56. The front elevation and profile views of a bridge panel are shown in Fig. 2-60. Scale: $\frac{1}{8}'' = 1'-0''$. What is the true length and slope of the diagonals AB and CD?

Ans. T.L. of $AB = 27'-0''$
Slope of $AB = 26°30'$
T.L. of $CD = 32'-0''$
Slope of $CD = 21°45'$

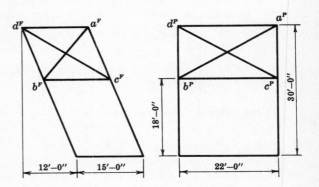

Fig. 2-60

57. A sewer line connecting four manholes is located as follows: The line from manhole A to manhole B has a bearing of N 45° E, is 120' long, and slopes downward at 10°; from manhole B to manhole C the line bears due east, is 100' long, and slopes downward at 15°; from manhole C to manhole D the line bears S 30° E, is 140' long, and slopes downward at 18°. Scale: 1" = 40'. If a sewer line were connected directly from manhole A to manhole D what would be its true length, slope, and bearing?

Ans. T.L. $AD = 264'$
Slope $AD = -20°$
Bearing $AD = $ S 82°30' E

58. A flagpole, DE, extends 10' into the air. Three guy wires are attached to the pole 8'' below the top and are fastened at various levels to anchors as shown in Fig. 2-61. Scale: $\frac{1}{4}'' = 1'-0''$. What is the true length, slope, and bearing of each guy wire?

Ans.

Line	True Length	Slope	Bearing
AD	$11'-4''$	$55°$	N 53° W
BD	$11'-10''$	$44°30'$	N 45° E
CD	$9'-11''$	$47°$	S 17° W

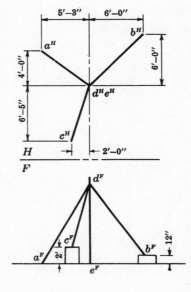

Fig. 2-61

59. Two pipes, AB and CD, intersect at D as shown in Fig. 2-62. They both slope downward towards D and the pipe from A has a slope of 45°. If point C is 10' lower in elevation than point A, what is the true length and slope of CD? What is the true length of pipe AB? Show both pipes in the front elevation view. Scale: 1" = 50'.

Ans. T.L. $CD = 142'$
Slope $CD = 39°$
T.L. $AB = 198'$

60. Three structural members AD, BD, and CD, are attached to a vertical wall. B is 8' east of A and 6' below A. Point C is 14' due east of A and 5' above A. Point D is 6' east, 5' south of A and 4' above A. Scale: $\frac{1}{4}'' = 1'-0''$. Determine the angle each structural member makes with the wall. What is the true length of each structural member?

Ans. T.L. $AD = 8'-9\frac{1}{2}''$
Angle with wall $= 35°30'$

T.L. $BD = 11'-5''$
Angle with wall $= 26°45'$

T.L. $CD = 9'-6''$
Angle with wall $= 32°$

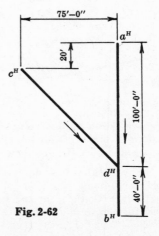

Fig. 2-62

61. Fig. 2-63 below shows the plan and front elevation views of a construction elevator support frame. The height of the frame is 30′ and has four guy wires attached as shown. Scale: $1'' = 20'$. What is the true length, slope, and bearing of wires A and C? What is the true length and slope of wires B and D? What angles do the wires B and D make with the structural support to which they are attached?

Ans. Line	T.L.	Slope	Bearing	Angle with Support
A	36′–6″	41°30′	N 42° W	
B	38′–0″	45°		45°
C	44′–0″	28°30′	N 67°30′ E	
D	39′–6″	37°		53°

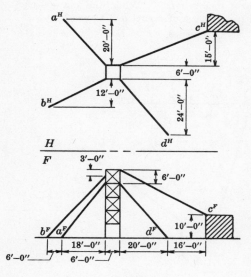

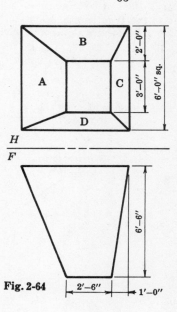

Fig. 2-63

Fig. 2-64

62. Fig. 2-64 above shows a metal transition piece for a fume exhaust system. Scale: $\frac{3}{8}'' = 1'-0''$. Show the true size of planes A, B, C, and D. If the inside of the duct is to receive a protective coating of epoxy resin paint, how many square feet of metal will receive the paint coat? (Disregard metal thickness.)   *Ans.* 119 ft²

63. A sewer line under Side Street slopes downward at -15% until it reaches another sewer line under Main Street which slopes downward at -30%. See Fig. 2-65. What is the true size of the angle between the two sewer lines? If a sewer line sloping -25% under Cross Street intersected the Main Street sewer line, what angle would be formed by these two intersecting lines?

Ans. Angle between Side St. and Main St. = 137°
Angle between Cross St. and Main St. = 94°

64. A flat metal plate, $ABCD$, is located as follows: B is 3′ west, 6′ north of A and 6′ below A; C is 8′ east, 4′ north of A and 4′ above A; D is 3′ east and 7′ north of A. Scale: $\frac{1}{4}'' = 1'-0''$. Locate point D in the front view. Determine the true size and slope of the plane. A point X on the plane is located 5′ above B and 8′ away from C. If X is the center of a circle, determine diameter of the largest circular hole which can be cut in the plate and yet maintain a 6″ edge clearance. Show the hole in all views.

Ans. Slope = 45°, Diameter = 5′–6″

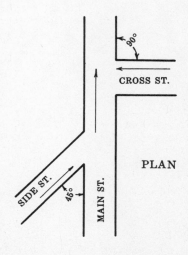

Fig. 2-65

65. Fig. 2-66 below shows the front elevation and profile views of an airplane landing gear. Scale: $\frac{1}{2}'' = 1'-0''$. Determine the true length of the landing gear supports AB, AC, and AD.
    Ans. T.L. $AB = 7'-5''$, T.L. $AC = 5'-4''$, T.L. $AD = 5'-9\frac{1}{2}''$

66. An airplane at A, elevation 1500', is bearing S 45° W at a speedometer reading of 120 miles per hour. It is gaining altitude at the rate of 2100' per minute. From A, a ship is sighted at bearing S 10° E on a 40° angle of depression. Twenty seconds later the ship is sighted at bearing N 60° E on a 45° angle of depression. Scale: 1'' = 1000'. Assuming the ship's course to be steady, what is its bearing and speed in knots? (1 Knot = 6080'/hr.)   *Ans.* Bearing = N 63° W, Speed = Approx. 28 Knots

67. A truck hoist is constructed as shown in Fig. 2-67 below. Point A is located at the mid-point of ED. AB has a slope of 60° and a bearing of N 15° E. Scale: $\frac{1}{4}'' = 1'-0''$. Determine the true length of AB. What is the true length, slope, and bearing of structural member DG? When the main structural member ED appears as a point, what angles are formed by each set of structural braces?
    Ans. T.L. $AB = 3'-5''$, T.L. $DG = 8'-0''$, Slope $DG = 48°$, Bearing $DG = $ N 41° W, Angle $BAC =$
        60°30', Angle $FDG = 98°$

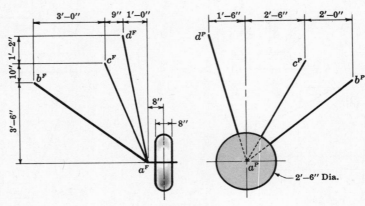

Fig. 2-66

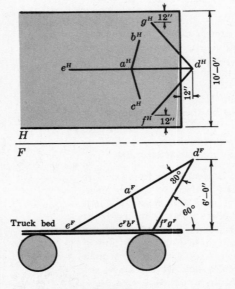

Fig. 2-67

Chapter 3

Skew Lines

THEOREMS

A theorem is a true statement capable of being proved.

The following are a group of theorems which the student of Descriptive Geometry should be able to recognize as self-evident truths. This can only be done by careful study and analysis. The truths expressed in these theorems will greatly aid in the selection of methods involved in the solution of problems contained in the next several chapters.

LINES and PLANES

(1) Two lines perpendicular to the same plane are parallel.

(2) Two lines lying on the same plane must either be parallel to each other or will intersect each other.

(3) Two lines parallel to each other in space will appear parallel, or as points, in all orthographic views.

(4) Two lines appearing parallel to each other in one view are not necessarily parallel to each other in space.

(5) Two perpendicular lines will appear perpendicular in any view that shows a normal view of either or both of the lines. (These lines need not intersect.)

(6) A line external to a plane and parallel to any line on the plane is also parallel to the plane itself.

(7) A line perpendicular to a plane is perpendicular to every line on that plane. An end view of the line shows the true length of every line on the plane.

(8) If a line is perpendicular to one of two perpendicular planes, it is parallel to the other.

(9) If a line is parallel to two intersecting planes, it is parallel to their intersection.

(10) If a line is perpendicular to a plane, then every plane through that line is also perpendicular to that plane.

(11) A point on a plane may have an infinite number of lines passing through it. However, only one of these lines can have the same slope as the plane and this line must be perpendicular to a level line on the plane.

(12) No line on a plane can have a slope greater than the slope of the plane.

(13) A plane angle may project smaller or larger than its true size.

(14) A normal view of a plane is an end view of every line perpendicular to that plane.

(15) An edge view of a plane is a normal view of every line perpendicular to that plane.

(16) If two parallel planes are cut by a third plane, the intersections are parallel.

(17) If a plane is perpendicular to two intersecting planes, it is perpendicular to their intersection.

(18) If two planes are perpendicular, a line in one of them perpendicular to their intersection is perpendicular to the other plane.

(*19*) The angle a line makes with a plane is the angle the line makes with its projection on that plane.

(*20*) If two planes are parallel, any line in one plane is also parallel *to the other plane*.

SKEW LINES

Skew lines are lines which are non-intersecting and non-parallel.

The principles involved in the study of skew line relationships are very important and should be thoroughly understood by every student of engineering and science. This chapter reveals the practicality of studying these basic principles and their application to several fields of endeavor.

One of the most common engineering problems is that of determining the true length, slope, and bearing of the shortest line connecting two skew lines. Problems which involve clearances between cables, pipes, braces, etc. are typical of those encountered in the engineering profession.

Another application of these principles can be seen in problems which require the shortest connecting tunnel between two mine tunnels. Again the perpendicular distance will be the required distance. If the connecting tunnel is to have a specific grade, a simple principle must be followed which would apply to all similar problems.

Still another practical application of the principles explained in this chapter involves finding the shortest connection between two oblique sewer lines or pipelines (see Fig. 3-1). In this case the shortest possible sewer or pipe connection will probably involve the use of right-angled tees since the shortest distance between any two skew lines is a third line which is perpendicular to both.

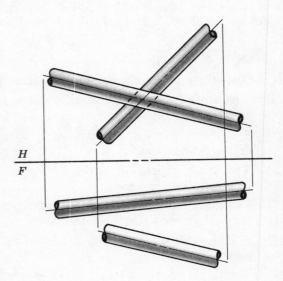

Fig. 3-1. Non-intersecting, Non-parallel Pipelines

3.1 To DRAW a PLANE CONTAINING ONE GIVEN LINE and PARALLEL to ANOTHER GIVEN LINE

Analysis: A geometric theorem tells us that if a line external to a plane is parallel to a line on the plane, then both the external line and the plane are parallel. Therefore through one of the given lines a line should be drawn which is parallel to the other given line. The two lines which now intersect determine a plane having a line parallel to a given line in space. The required plane must be represented by the two intersecting lines in at least two views. It must be remembered that intersecting lines can represent a plane only when the intersection can be projected between related views as a common point.

Example: In Fig. 3-2 below we have given the plan and front elevation views of two non-intersecting, non-parallel lines *AB* and *CD*. Through any point on the line *CD*, in this case *Z*, the line *XY* is drawn parallel to the line *AB* in space. This procedure is followed in both views, thus determining a plane *CDXY*, one line of which, *XY*, is parallel to the line *AB* in space.

This solution can be verified by drawing another view showing the plane as an edge. In this view the line *AB* will appear parallel to the plane.

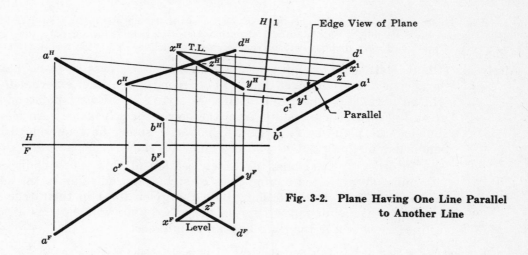

Fig. 3-2. Plane Having One Line Parallel
to Another Line

3.2 To DETERMINE the SHORTEST DISTANCE BETWEEN any TWO NON-INTERSECTING, NON-PARALLEL LINES

Analysis: **(Line Method)** The shortest distance between two non-intersecting, non-parallel lines will be the perpendicular distance. This perpendicular distance can be located at only one possible position in space. In a view showing one of the lines in its true length, the shortest distance will appear perpendicular to it. However, if either skew line appears as a point, the shortest distance will be perpendicular to the other skew line. In this same view, the shortest distance will be in its true length. Therefore the solution of the problem depends upon showing one of the skew lines as a point and projecting a perpendicular from this point view of a line to the other skew line.

Example: The two non-intersecting, non-parallel lines *AB* and *CD* are given in both the plan and front elevation views in Fig. 3-3 below. Since either of the lines can be shown in its true length, draw an auxiliary elevation view in which the true length of the line *AB* will appear. Also show line *CD* in this view. The shortest distance will be perpendicular to line *AB* but its exact location is yet unknown. Draw an inclined view showing the line *AB* as a point. The shortest distance will be the perpendicular distance from *AB* to the line *CD* even though the true length of *CD* does not appear in this view. Label the shortest distance line *XY* and project this common perpendicular back to view 1. Since *XY* is in its true length in view 2, it must be parallel to the folding line 1-2 in view 1. Also *XY* must be perpendicular to *AB* in view 1, where *AB* is shown in true length. Therefore point *X* can be located in view 1. Project *XY* to the *H* and *F* views.

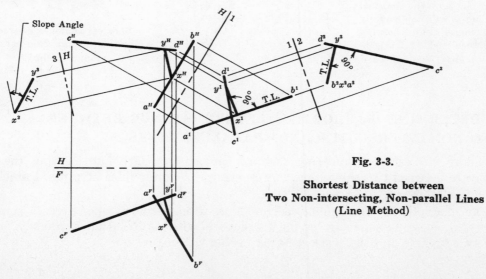

Fig. 3-3.

Shortest Distance between
Two Non-intersecting, Non-parallel Lines
(Line Method)

The bearing of the shortest distance line can be measured on the plan view. To determine the slope of the shortest distance line, a new auxiliary elevation view is drawn showing only the line XY in its true length. This true length distance will of course check with line XY in the inclined view 2.

Analysis: (Plane Method) Through one of the given lines pass a plane having a line parallel to the other given line. This second given line, external to the plane, will be parallel to the plane itself (see Art. 3.1). Every point on the second given line will be equidistant from the newly-formed plane. An elevation view showing the plane as an edge will not only show the true length of the perpendicular distance but it will also show its slope.

In order to determine the exact location of this shortest distance, a new view must be projected having lines of sight perpendicular to the edge view of the plane. This view will show both the given lines in their true length and the perpendicular distance between these two lines will show as a point in this inclined view where the lines appear to intersect.

Example: The plan and front elevation views of lines AB and CD are given in Fig. 3-4 below. A plane CDE is constructed containing line CD and having DE parallel to line AB. Draw an auxiliary elevation view showing the plane as an edge. The line AB will be parallel to the edge view of the plane in this view. The perpendicular distance between the plane and the line will be the true length of the shortest distance.

To determine the exact location of this shortest distance, project an inclined view directly off the auxiliary elevation view to see the true length of both lines AB and CD. Since folding line 1-2 was placed parallel to the edge view of the plane, the inclined view will also show the true size of the plane. The point view of the shortest distance XY will be located where the true length of both lines appear to intersect. Project line XY back to the auxiliary elevation view and thence on to the other views. Note that the plan view of line XY will be parallel to the H-1 folding line, since the true length of XY is shown in view 1.

The slope of the shortest distance line will appear in view 1, whereas the bearing will be determined by its position in the plan view.

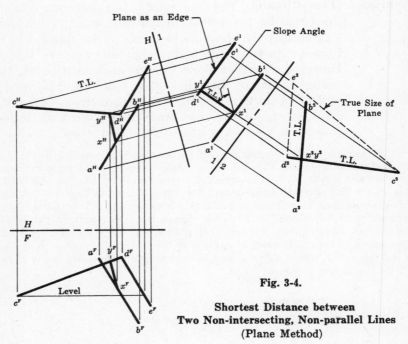

Fig. 3-4.

**Shortest Distance between
Two Non-intersecting, Non-parallel Lines
(Plane Method)**

3.3 To DETERMINE the SHORTEST LEVEL DISTANCE BETWEEN TWO NON-INTERSECTING, NON-PARALLEL LINES

In many phases of engineering the problem arises relative to locating the shortest level distance between two skew lines representing the center lines of pipes, tunnels, wires, braces, etc.

As an example, in mining engineering it is often deemed imperative to connect two tunnels with a new passageway having no slope at all. Quite often a track is to be laid between two tunnels and the track is to be perfectly level.

Analysis: The shortest level distance between two skew lines will appear in its true length when the plane which contains one skew line and is parallel to the other appears as an edge. Even though the true length of the shortest level distance is known, its exact location is not determined in the auxiliary elevation view. Since the true length of the shortest level distance will appear in an elevation view, the point view of the shortest level distance will also appear in the elevation view having lines of sight parallel to the shortest level distance. Where the two given lines appear to intersect in this new elevation view, the shortest level distance line will appear as a point.

Note: The line method of Art. 3.2 cannot be used to determine the shortest level distance, because in the line method the view which shows the required line in its true length also shows an inclined line as a point and, therefore, the view cannot be an elevation view.

Example: In Fig. 3-5 below, the two skew lines *AB* and *CD* are given in both the plan and front elevation views. The plane *CDE* is drawn containing the line *DE* parallel to line *AB*. The auxiliary elevation view is drawn showing the plane as an edge and the line *AB* parallel to it. The shortest level distance is the level distance between the two parallel lines, one of which represents the edge view of a plane. This level distance will be parallel to the *H*-1 folding line. In order to show the point view of the shortest level distance, folding line 1-2 must be located perpendicular to the *H*-1 folding line. The elevation view 2 will determine the exact location of the level line as the point where the lines *AB* and *CD* appear to intersect. The level line *XY* is then projected back to the other views, and it will be noted that line *XY* in the plan view will appear in its true length. In the front elevation view, the line *XY* must be parallel to the *H-F* folding line, since every level line is parallel to the horizontal image plane.

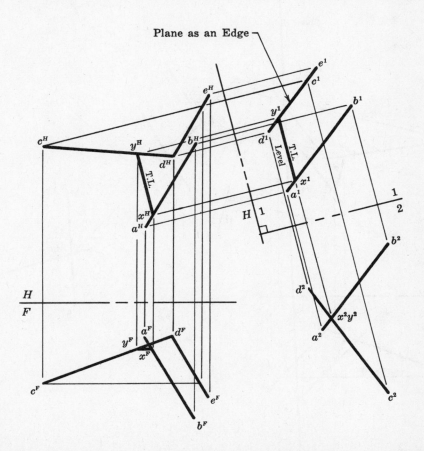

Fig. 3-5.   Shortest Level Distance between Two Non-intersecting, Non-parallel Lines

3.4 To DETERMINE the SHORTEST LINE of GIVEN SLOPE CONNECTING TWO NON-INTERSECTING, NON-PARALLEL LINES

Analysis: Again, if a plane is drawn containing one of the skew lines and parallel to the other skew line, an auxiliary elevation view showing the plane as an edge will also show the other skew line parallel to it. Since the plane and line appear parallel in an elevation view, the required line can be drawn at the specified slope angle. However, even though the direction of the grade line is determined, its exact location is not known. The projection of a new view, having the folding line perpendicular to the direction of the shortest line of given slope, will show the required line as a point where the skew lines appear to intersect.

Note: The line method of Article 3.2 cannot be used to determine the shortest line of given slope because in the line method the view which shows the required line in its true length also shows an inclined line as a point and therefore the view cannot be an elevation view.

Example: In Fig. 3-6 below, the two skew lines *AB* and *CD* are given in both the plan and front elevation views. The plane *CDE* is drawn containing the line *DE* parallel to line *AB*. The auxiliary elevation view is drawn showing the plane as an edge and the line *AB* parallel to it. Any shortest grade line desired will appear in its true length in this view and will make the specified slope angle with the folding line *H*-1. To obtain the exact location of the required line of given slope, draw folding line 1-2 perpendicular to the directional line having the given slope, and the intersection point of *AB* and *CD* in this view shows the required line as a point. Simple projection will locate the line in the other views.

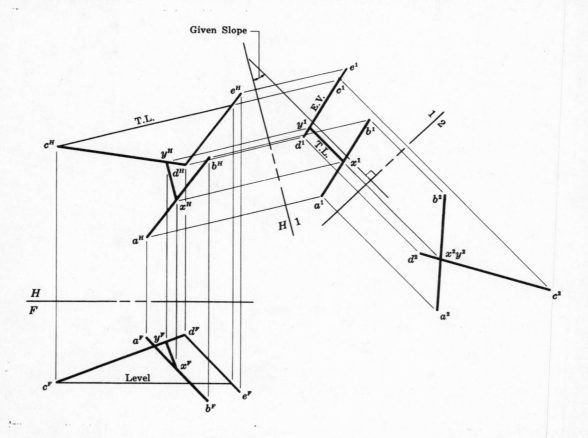

Fig. 3-6. Shortest Line of Given Slope between Two Non-intersecting, Non-parallel Lines

Solved Problems

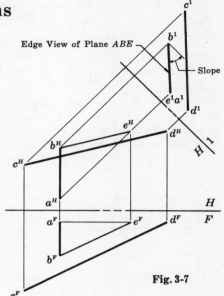

Fig. 3-7

1. **Given:** Point B is located $1'-6''$ due north of A and $1'$ below A. Point C is located $1'$ north, $1'$ west of A and $2'$ below A. Point D is $3'$ east, $2'$ north of A and at the same elevation as A. See Fig. 3-7. Scale: $\frac{3}{8}'' = 1'-0''$.

   Problem: Using the two skew lines AB and CD, draw a plane which contains the line AB and which is parallel to the line CD. Draw an auxiliary elevation view to verify your solution. Determine the slope of the plane.

   Solution:

   In both the plan and front elevation views, draw line BE parallel to the line CD. The line CD is now parallel to the plane ABE. Draw auxiliary elevation view 1 to show the line and plane parallel to each other. The slope is measured in view 1.

   Ans. Slope $= 44°30'$

2. **Given:** Plan view of plane ABC and line XY. Front elevation view of plane ABC and point X. Line XY is parallel to the plane. Refer to Fig. 3-8.

   Problem: Complete the front view of line XY.

   Solution:

   From point B in the plan view draw a line parallel to the plan view of XY. It will intersect the line AC at point D. Project point D down to the front view and connect it to point B. From point Y in the plan view project a line down to the front view. From point X in the front view draw a line parallel to BD in the front view. Where this line intersects the projection from Y in the plan view will determine the position of Y in the front view.

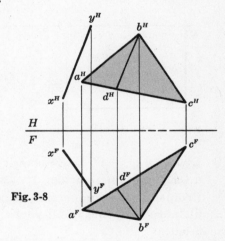

Fig. 3-8

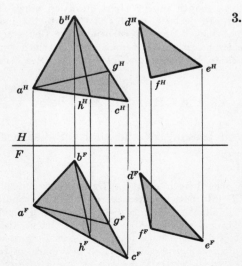

Fig. 3-9

3. **Given:** Plan view of planes ABC and DEF. Front elevation view of plane ABC and point F. Planes ABC and DEF are parallel. See Fig. 3-9.

   Problem: Complete the front elevation view of plane DEF.

   Solution:

   In the plan view of the plane ABC draw a line from point A parallel to the plan view of the line FE. From point B in the plan view draw a line parallel to the plan view of the line DF. Project these newly drawn lines to the front view of plane ABC. From point F in the front view draw a line parallel to AG in the front view until it intersects the projection of E from the plan view. Again from point F in the front view draw a line parallel to BH in the front view until it intersects the projection of D from the plan view. Connect points D, E, and F to obtain the front view of plane DEF parallel to plane ABC.

4. **Given:** Two lines *AB* and *CD*. Point *B* is located 1′−6″ due north of *A* and 2′ below *A*. Point *C* is 2′−6″ east, 2′ north of *A* and 1′ below *A*. Point *D* is located 2′−6″ east, 6″ south of *A* and 2′ below *A*. See Fig. 3-10 below. Scale: $\frac{1}{2}''=1'-0''$.

Problem: Determine the shortest distance between the lines *AB* and *CD*. What is the slope and bearing of the shortest distance?

Solution:

Using the given data, draw the plan and front elevation views of lines *AB* and *CD*. It will be noted that the two lines are parallel profile lines. A profile view will show the shortest distance as a point. The shortest distance will be a frontal-level line. Label the line *XY* and show it in all views. The true length of the shortest distance will appear in both the plan and front elevation views. *Ans.* T.L. = 2′−6″, Slope = 0°, Bearing = Due East-West

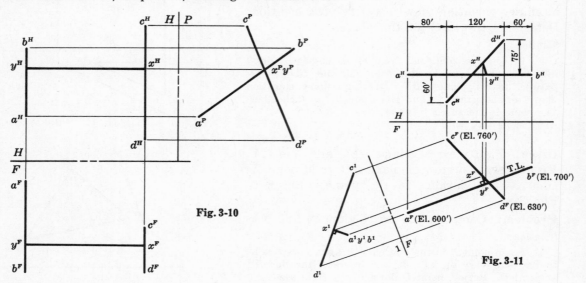

Fig. 3-10   Fig. 3-11

5. **Given:** Fig. 3-11 above shows the plan and front elevation views of a power line and a telephone line. Scale: 1″ = 200′.

Problem: Determine the clearance between the two lines. Represent the clearance distance by a line in all views. Use the line method.

Solution:

Since line *AB* is a frontal line, it will appear true length in the front elevation view. Place folding line *F*-1 perpendicular to the true length of *AB*. Project lines *AB* and *CD* into the inclined view 1. Line *AB* will appear as a point in this inclined view. Draw a perpendicular line from the point view of *AB* to line *CD*. This perpendicular line from *AB* to *CD* represents the true clearance between the two lines. Project this clearance line back to the front and plan views. See Art. 3.2 and Fig. 3-2.   *Ans.* Clearance = 30′

6. **Given:** Two pipelines are represented by lines *AB* and *CD* in Fig. 3-12 below. Scale: 1″ = 30′.

Problem: Determine the true length, slope, and bearing of the shortest distance between the two pipelines.

Solution:

Draw an auxiliary elevation view showing the true length of *AB*. Draw an inclined view showing the point view of the line *AB*. Project line *CD* into both of these views. The perpendicular distance from the point view of *AB* to the line *CD* is the shortest distance between the two pipelines. Since view 2 shows the true length of the shortest distance, view 1 will show the shortest distance to be parallel to the folding line 1-2. Project the shortest distance back to the plan view from which the bearing can be determined. An auxiliary elevation view having folding line *H*-3 parallel to the plan view of the shortest distance will reveal both the true length and slope of the required line. (See Art. 3.2.)   *Ans.* T.L. = 20′, Slope = 57°30′, Bearing = N 6° E

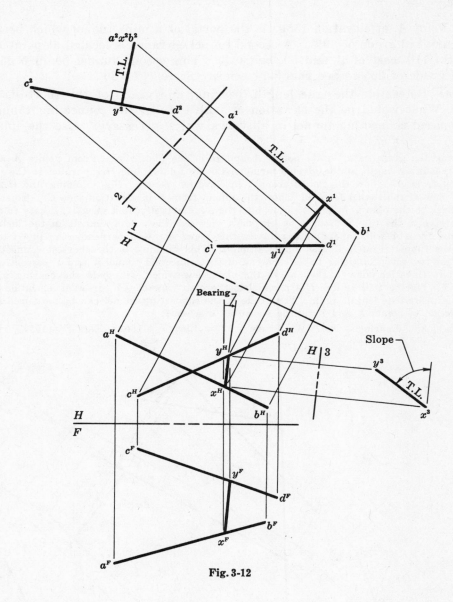

Fig. 3-12

7. **Given:** A 3″ cube. Scale: 3″ = 1′−0″.

Problem: What would be the true length and slope of the shortest distance between the non-intersecting diagonals of any two adjacent faces? Use the line method. Refer to Fig. 3-13.

Solution:

Draw the plan and front elevation views of the cube showing the two non-intersecting diagonals. Draw an auxiliary elevation view 1 to show the horizontal diagonal as a point. Show the diagonal on the vertical surface of the cube in this view also. Draw a perpendicular line from the point view of one diagonal to the other diagonal. This perpendicular line represents the true length of the shortest distance between the two non-intersecting diagonals. The slope of the shortest distance can also be measured in the auxiliary elevation view 1. *Ans.* T.L. = $1\frac{3}{4}$″, Slope = 35°

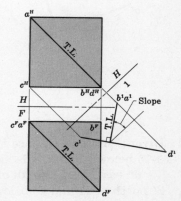

Fig. 3-13

8. **Given:** Point A, at elevation 1550′, is the portal of a mine tunnel which bears N 60° E on a downward grade of 30%. A second tunnel entrance is located at point B which is 50′ south, 110′ east of A and 50′ below A. This second tunnel bears N 45° E on an upward grade of 35%. See Fig. 3-14 below. Scale: 1″ = 80′.

Problem: Determine the true length, bearing, and grade of the shortest connecting tunnel. What would be the elevation at both ends of the connecting tunnel? Show this proposed connecting tunnel in all views where necessary. Use the line method.

Solution:

   Using the given data, locate points A and B in the plan view. From points A and B, draw lines of indefinite length at the given bearings. Locate folding line H-1 parallel to the plan view of the tunnel from A. Show the tunnel from A in elevation view 1. Place folding line H-2 parallel to the plan view of the tunnel from B. Show the tunnel from B in elevation view 2. Project the tunnel from B back to the plan view and then on into the auxiliary elevation view 1. Locate folding line 1-3 perpendicular to the tunnel from A in elevation view 1. Show both tunnels in the inclined view 3. The tunnel from A will appear as a point in this view. Draw a perpendicular line from the point view of the tunnel from A to the tunnel from B. This distance is the shortest connecting distance between the two given tunnels. Label each end of the connecting tunnel X and Y respectively. Project XY back to elevation view 1, then on to the plan view which will yield the bearing of connecting tunnel XY. Placing folding line H-4 parallel to the plan view of XY, draw an elevation view of XY showing it in its true length again. The grade of the connecting tunnel can be measured in this view. The elevation of points X and Y can be measured in view 1.

Ans. T.L. = 80′, Bearing = N 37° W, Grade = 42%, Elev. X = 1503′, Elev. Y = 1534′

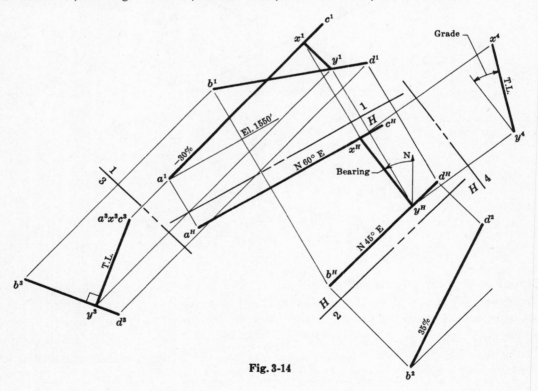

Fig. 3-14

9. **Given:** AB and CD are the centerlines of two 3″ diameter pipe lines as shown in Fig. 3-15 below. Point B is located 2′ east, 3′ north of A and 2′ below A. Point C is located 1′−6″ east, 1′−6″ north of A and at the same elevation as A. Point D is located 1′ west, 3′ north of A and 2′ below A. Scale: $\frac{1}{2}$″ = 1′−0″.

Problem: How much clearance, if any, is there between the two pipes? Use the line method.

Solution:

Using the given data, draw the plan and front elevation views of lines AB and CD. Draw an auxiliary elevation view 1 showing both pipes, one of which, CD, appears in its true length. Draw an inclined view 2 showing pipe centerline CD as a point. Show pipe centerline AB in this view also. Draw a perpendicular line from the point view of CD to the line AB. This perpendicular distance, minus the pipe itself, determines the clearance between the two pipes.   *Ans.* Clearance $= 4\frac{1}{2}''$

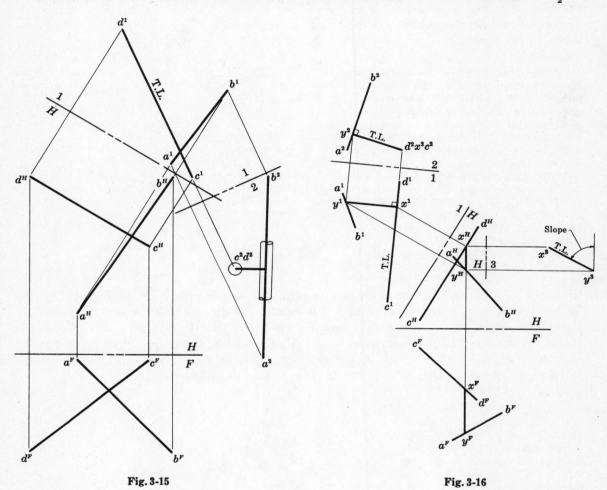

Fig. 3-15   Fig. 3-16

10. **Given:** Two mining shafts AB and CD in Fig. 3-16 above. Point B is 300' south, 260' east of A and 150' above A. Point C is located 360' south, 175' west of A and 500' above A. Point D is 160' north, 130' east of A and 220' above A. Scale: $1'' = 500'$.

Problem: It is proposed to connect these two shafts with a new ventilating tunnel. Determine the true length, slope, and bearing of the shortest possible ventilating tunnel connecting the shafts. Show this ventilating tunnel in all views. Use the line method.

Solution:

Draw the plan and front elevation views of shafts AB and CD. Draw an auxiliary elevation view 1 showing the true length of CD. Project AB into this view also. Draw an inclined view 2 which will show the point view of shaft CD. Project AB into this view and draw a perpendicular line from the point view of CD to the line AB. This line, XY, will be the shortest distance between the two shafts. Project XY back to the plan view to determine the bearing of the ventilating tunnel. Draw auxiliary elevation view 3 to obtain the true length and slope of XY. The true length of XY in both views 2 and 3 must be the same.

Ans. T.L. $= 275'$, Slope $= 61°$, Bearing $=$ Due North-South

11. **Given:** Two skew lines, AB and CD in Fig. 3-17. Point B is located 2'–6'' due east of A and 1' below A. Point C is located 1' east, 1' south of A and 1' above A. Point D is located 3' east, 1' north of A and 1'–6'' above A. Scale: $\frac{3}{8}'' = 1'$–$0''$.

Problem: Determine the true length, slope, and bearing of the shortest distance from line AB to line CD. Show the shortest distance in all views.

Solution:

Draw the plan and front elevation views of lines AB and CD. Draw an inclined view 1 which will show line CD and the point view of line AB. From the point view of line AB, draw a line perpendicular to CD. Label this line XY. The line XY in this view is the shortest distance between lines AB and CD. Project XY back to the other views. The bearing of the shortest distance from AB to CD will appear in the plan view. Draw an auxiliary elevation view 2 to obtain the slope of XY as well as a check on the true length obtained in view 1.

Ans. T.L. = $1'$–$7\frac{1}{3}''$, Slope = $52°$, Bearing = S $32°$ E

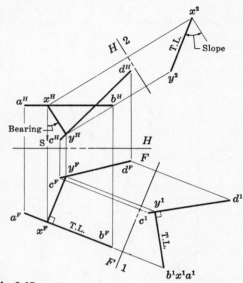

Fig. 3-17

12. **Given:** Lines AB and CD are the centerlines of two sewer pipes. Point B is 15' north, 75' west of A and 55' above A. Point C is located 25' due south of A and 75' above A. Point D is 40' north, 40' west of A and 40' above A. Scale: 1'' = 80'. See Fig. 3-18.

Problem: Using only 90° tees, show where to connect the two sewer pipes with a third pipe of shortest possible length. Determine the true length, slope, and bearing of the third pipe. Show this third pipe in all views. Use the plane method.

Solution:

Using the given data, draw the plan and front elevation views of lines AB and CD. Construct a plane, CDE, containing the line CE parallel to the line AB in both views. Draw an auxiliary elevation view 1 to obtain an edge view of the plane parallel to line AB. The perpendicular distance between AB and CD in view 1 is the true length of the shortest connecting pipe. To obtain the exact location of this shortest distance, an inclined view 2 must be drawn to show the connecting pipe as a point. Label the point view of the connecting pipe XY and project it back to the other views. The bearing of the third pipe will be measured in the plan view. Auxiliary elevation view 1 will yield both the true length and slope of the shortest connecting pipe.   *Ans.* T.L. = $26'$–$0''$, Slope = $35°$, Bearing = N $41°$ E

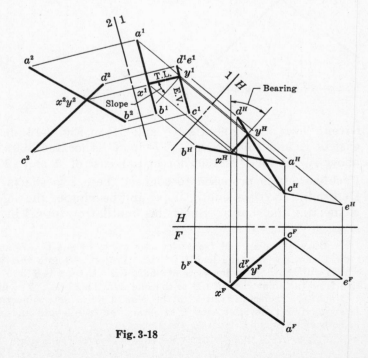

Fig. 3-18

13. Given: Two mining tunnels *AB* and *CD*. Point *B* is 360′ east, 310′ south of *A* and 400′ below *A*. Point *C* is 60′ east, 400′ south of *A* and 320′ below *A*. Point *D* is 375′ east, 60′ south of *A* and 50′ below *A*. Refer to Fig. 3-19. Scale: 1″ = 400′.

Problem: What would be the true length and bearing of the shortest level tunnel connecting *AB* and *CD*? Show the shortest level tunnel in all views.

Solution:

Locate tunnels *AB* and *CD* in the plan and front elevation views. In both views, draw a plane *CDE* containing the line *DE* parallel to *AB*. Draw an auxiliary elevation view 1 to show line *AB* parallel to the plane *CDE*. The level distance between the two parallel lines is the required distance but the exact location of this level tunnel is not known. A point view of the level tunnel will establish its location in view 1. Locate folding line 1-2 perpendicular to the *H*-1 folding line. View 2 will therefore be an elevation view which will show the level line as a point. Label the level tunnel *XY* and project it back to the other views. The true length will be measured in either the plan view or auxiliary elevation view 1. The bearing is obtained in the plan view. *Ans.* T.L. = 95′, Bearing = N 14° W

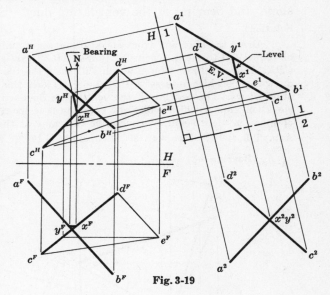

Fig. 3-19

14. Given: *AB* and *CD* are the centerlines of two partial pipelines (see Fig. 3-20). Point *B* is 15′ north, 60′ west of *A* and 20′ above *A*. Point *C* is 15′ south, 5′ east of *A* and 80′ above *A*. Point *D* is 60′ north, 25′ west of *A* and 30′ above *A*. Scale: 1″ = 80′.

Problem: Show where to connect the two pipelines with the shortest level pipe. What would be the true length and bearing of the shortest possible level pipe? Show the shortest level distance in all views. Extend the lines, if necessary, for the solution of the problem.

Solution:

Using the given data, draw the plan and front elevation views of lines *AB* and *CD*. Draw a plane *CDE* containing the line *DE* parallel to the line *AB* in both views. Draw an auxiliary elevation view 1 to show the edge view of the plane parallel to the line *AB*. The level distance between the two parallel lines in view 1 is the true length of the shortest possible level connecting pipe. To determine the exact location of the shortest level distance, draw elevation view 2 with folding line 1-2 perpendicular to folding line *H*-1. The intersection of lines *AB* and *CD* in view 2 will locate the point view of the shortest level distance. Label the intersection *XY* and project it back to the other views. The bearing of *XY* will be measured in the plan view and the level line *XY* must appear parallel to the *H-F* folding line when it is projected to the front view. *Ans.* T.L. = 68′, Bearing = N 32°30′ E

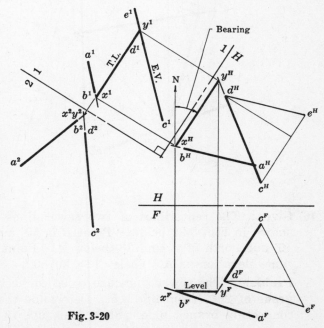

Fig. 3-20

15. Given: Two tunnels, *AB* and *CD* in Fig. 3-21 below.   Point *B* is located 80′ east, 20′ north of *A* and 30′ below *A*.   Point *C* is located 10′ east, 75′ north of *A* and 60′ above *A*. Point *D* is located 100′ east, 45′ north of *A* and 45′ above *A*.   Scale: 1″ = 60′.

Problem: The two tunnels are to be connected by a ventilating duct from the mid-point of *AB*, entering the second tunnel *CD* at an angle of 60°.   Determine the true length, slope, and bearing of the ventilating duct.   Show the duct in all views.

Solution:

Draw the plan and front elevation views of tunnels *AB* and *CD*. In both views, connect the ends of tunnel *CD* to the mid-point of tunnel *AB*. Draw an inclined view 1 to show the edge view of plane *CDX*. Draw inclined view 2 which will show the true size of plane *CDX*. From point *X* in this view, draw a line which intersects *CD* at 60°. Label this intersection point *Y* and project it back to the other views. The bearing of the ventilating duct will be measured in the plan view. Draw auxiliary elevation view 3 which will show both the true length and slope of line *XY*. The true length of *XY* obtained in view 3 will, of course, be the same as that which is obtained in view 2.

Ans.   T.L. = 97′, Slope = 49°30′, Bearing = N 17°30′ W

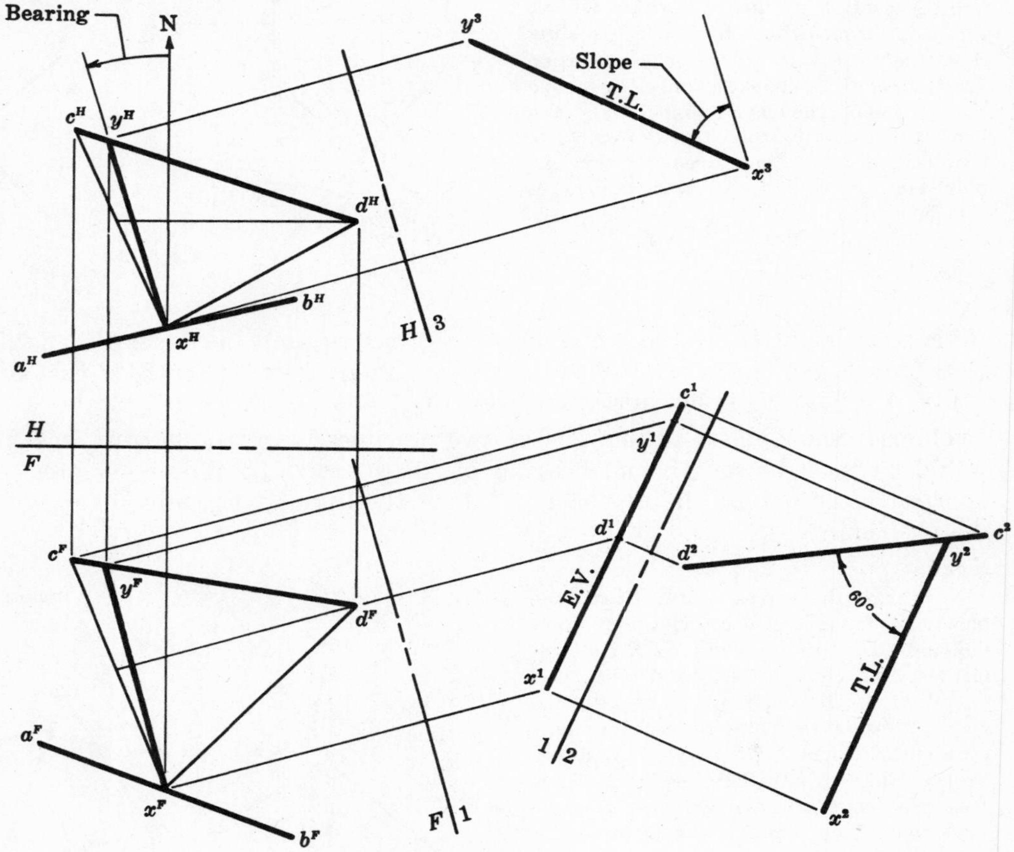

Fig. 3-21

16. Given: The centerlines of two sewer lines are determined by lines *AB* and *CD* as shown in Fig. 3-22 below.   Point *B* is 50′ due east of *A* and 20′ above *A*.   Point *C* is 10′ due north of *A* and 30′ below *A*.   Point *D* is 30′ east, 30′ south of *A* and at the same elevation as *A*.   Scale: 1″ = 30′-0″.

Problem: The sewer lines are to be connected by a branch pipe having a downward grade of 25% from line *AB*.   Determine the true length and bearing of the shortest connecting branch pipe.   Show this pipe in all views.

Solution:

Using the given data, draw the plan and front elevation views of lines *AB* and *CD*. Draw the plane *ABE* containing the line *AE* parallel to *CD* in both the plan and front views. Draw an auxiliary elevation view 1 showing line *CD* parallel to the edge view of plane *ABE*. To obtain the exact location of the required line of given slope, draw folding line 1-2 perpendicular to the directional line having the grade of −25%. The intersection of lines *AB* and *CD* in the inclined view 2 will show the required line as a point. Label the connecting pipe *XY* and project it back to the other views. The true length appears in auxiliary elevation view 1 and the bearing is measured in the plan view.

Ans. T.L. = 30′–6″, Bearing = S 40° E

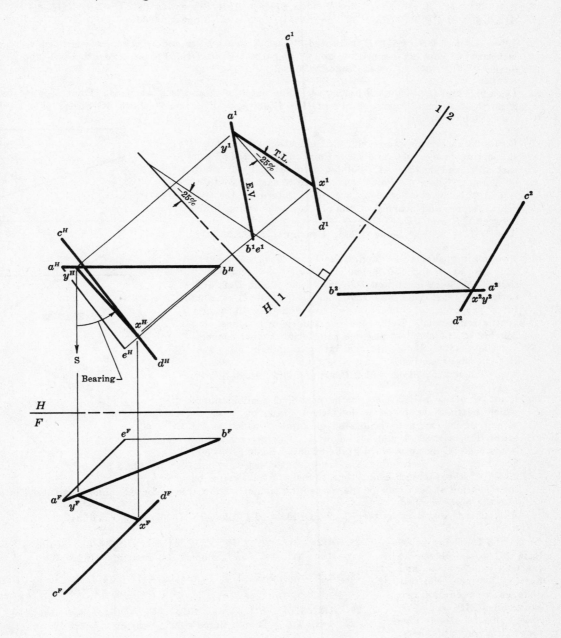

Fig. 3-22

Supplementary Problems

In each of the following problems 17 to 23 draw a plane which contains the line *AB* and is parallel to the line *CD*. To verify your solution draw an edge view of the plane to determine if the line *CD* lies parallel to the plane. Find the slope of the plane.   See Art. 1.7 for the coordinate system of problem layout.

17. $A(1, 1, 6)$ $B(3, 2\frac{1}{2}, 6)$ $C(\frac{1}{2}, 2, 4)$ $D(2\frac{1}{2}, 2, 5)$.   *Ans.* Slope = 60°
18. $A(1, \frac{1}{2}, 4)$ $B(2, 3, 6)$ $C(3, 1, 5\frac{1}{2})$ $D(4, 3, 5)$.   *Ans.* Slope = 65°
19. $A(2, 1, 5\frac{1}{2})$ $B(3, 3\frac{1}{2}, 4\frac{1}{2})$ $C(1, 2, 5)$ $D(4, 2, 5)$.   *Ans.* Slope = 67°
20. $A(1, 1, 5)$ $B(4, \frac{1}{2}, 5)$ $C(2, 2, 4)$ $D(4\frac{1}{2}, 3, 5\frac{1}{2})$.   *Ans.* Slope = 44°30′
21. $A(2, 3, 4)$ $B(4, 2, 6)$ $C(2, 1, 5)$ $D(5, 2\frac{1}{2}, 4)$.   *Ans.* Slope = 38°30′

22.   Lines *AB* and *CD*. Point *B* is located 3′ east, 3′ south of *A* and at the same elevation as *A*. Point *C* is located 2′ east, 2′ north of *A* and 3′ above *A*. Point *D* is 6′ east, 1′ south of *A* and 1′ above *A*. Scale: $\frac{1}{2}″ = 1′-0″$.   *Ans.* Slope = 70°30′

23.   Lines *AB* and *CD*. Point *B* is located 3′ due north of *A* and is 2′ above *A*. Point *C* is located 2′ east, 2′ north of *A* and 2′ above *A*. Point *D* is 4′ due east of *A* and 3′ above *A*. Scale: $\frac{1}{2}″ = 1′-0″$. *Ans.* Slope = 54°30′

24.   Fig. 3-23 shows the plan and front elevation views of two supporting braces, *AB* and *CD*. Point *B* is located 7′ due east of *A* and 4′ below *A*. Point *C* is 6″ east, 1′-6″ south of *A* and 2′ below *A*. Point *D* is located 8′ east, 4′ south of *A* and 2′ below *A*. Scale: $\frac{1}{2}″ = 1′-0″$. Determine the true length, slope, and bearing of the shortest connecting brace between *AB* and *CD*.
Ans. T.L. = 2′-1″,   Slope = 30°,   Bearing = N 18°30′ E

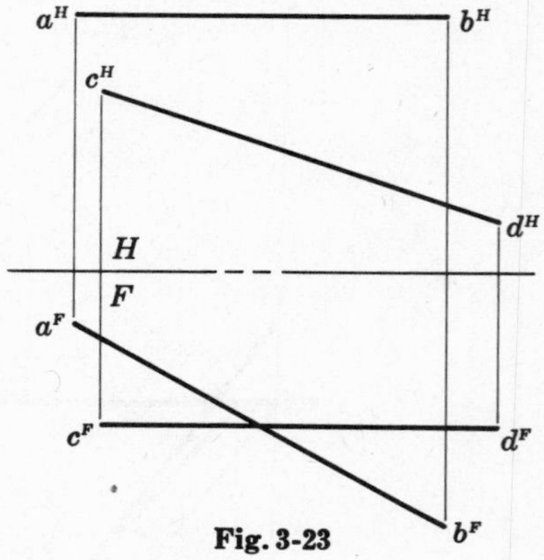

Fig. 3-23

25.   Two mining shafts *AB* and *CD*. Point *B* is 300′ south, 260′ east of *A* and 150′ above *A*. Point *C* is located 360′ south, 175′ west of *A* and 500′ above *A*. Point *D* is 160′ north, 130′ east of *A* and 220′ above *A*. Scale: 1″ = 200′. It is proposed to connect these two shafts with a new ventilating tunnel. Determine the true length, slope, and bearing of the shortest possible ventilating tunnel connecting the shafts. Show this ventilating tunnel in all views. Use the plane method.
Ans. T.L. = 275′, Slope = 61°, Bearing = Due North-South

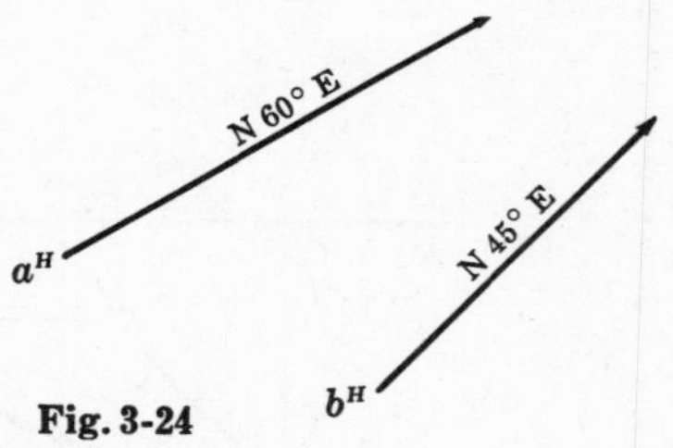

26.   Point *A*, at elevation 1550′, is the portal of a mine tunnel which bears N 60° E on a downward grade of 30%. A second tunnel entrance is located at point *B* which is 50′ south, 110′ east of *A* and 50′ below *A*. This second tunnel bears N 45° E on an upward grade of 35%. Scale: 1″ = 50′. See Fig. 3-24. Determine the true length, bearing, and grade of the shortest connecting tunnel. What would be the elevation at both ends of the connecting tunnel? Show this proposed connecting tunnel in all views. Use the plane method.
Ans. T.L. = 80′, Bearing = N 37° W, Grade = 42% Elev. *X* = 1503′, Elev. *Y* = 1534′

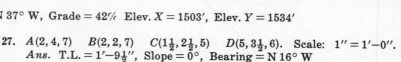

Fig. 3-24

In the following problems 27 to 31 determine the true length, slope, and bearing of the shortest distance between the two skew lines *AB* and *CD*. Show this shortest distance line in all views. Use the line method.

See Art. 1.7 for coordinate system of problem layout.

27. $A(2, 4, 7)$ $B(2, 2, 7)$ $C(1\frac{1}{2}, 2\frac{1}{2}, 5)$ $D(5, 3\frac{1}{2}, 6)$. Scale: 1″ = 1′-0″.
Ans. T.L. = 1′-9½″, Slope = 0°, Bearing = N 16° W

28. $A(2, 3, 5)$ $B(4\frac{1}{2}, 1, 6)$ $C(5\frac{1}{2}, 3\frac{1}{2}, 4\frac{1}{2})$ $D(4, 4, 6)$. Scale: 1″ = 1′-0″.
Ans. T.L. = 2′-2″, Slope = 54°, Bearing = N 63° E

29. $A(2, 3\frac{1}{2}, 5)$ $B(5, 3\frac{1}{2}, 6)$ $C(2, 4, 6\frac{1}{2})$ $D(5\frac{1}{2}, 1, 5\frac{1}{2})$. Scale: $\frac{1}{4}″ = 1′-0″$.
Ans. T.L. = 3′-6″, Slope = 34°, Bearing = N 19° W

30. $A(4, 1, 6)$ $B(7, 3, 5)$ $C(6, 1\frac{1}{2}, 7)$ $D(7, 1, 5)$. Scale: $\frac{1}{2}″ = 1′-0″$.
Ans. T.L. = 2′-7″, Slope = 42°, Bearing = N 52° E

31. $A(5, 1, 5)$ $B(7, 3, 5)$ $C(6, 1, 7)$ $D(7, 0, 5)$. Scale: $\frac{3}{8}″ = 1′-0″$.
Ans. T.L. = 4′-6″, Slope = 34°, Bearing = N 45° E

32. Lines *AB* and *CD* are the centerlines of two sewer pipes. Point *B* is 15′ north, 75′ west of *A* and 55′ above *A*. Point *C* is located 25′ due south of *A* and 75′ above *A*. Point *D* is 40′ north, 40′ west of *A* and 40′ above *A*. Scale: $1'' = 40'$. Using only 90° tees, show where to connect the two sewer pipes with a third pipe of shortest possible length. Determine the true length, slope, and bearing of the third pipe. Show this third pipe in all views. Use the line method.
Ans. T.L. = 26′−0″, Slope = 35°, Bearing = N 41° E

33. *AB* and *CD* are two control cables (see Fig. 3-25). The cables are located as follows: $A(1, 3, 7)$ $B(3, 4, 4\frac{1}{2})$ $C(\frac{1}{2}, 4, 4\frac{1}{2})$ $D(2\frac{1}{2}, 2, 6)$. Scale: $1'' = 1'-0''$. Using the line method, determine the clearance between the two cables. See Art. 1.7 for the coordinate system of problem layout. *Ans.* Clearance = 10″

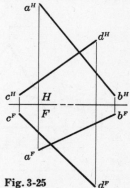

34. Points *A* and *B* are the anchors for two cables. Point *B* is 4′ south, 2′ east of *A* and 5′ below *A*. The bearing of the cable from *A* is S 60° E and it slopes downward 30°. The cable from *B* bears N 45° E and it has a slope of −20°. Scale: $\frac{1}{4}'' = 1'-0''$. Determine the clearance between the two cables. *Ans.* Clearance = 2′−10″

35. Two utility lines, *AB* and *CD*. Point *B* is located 5′ east, 30′ south of *A* and 15′ below *A*. Point *C* is located 15′ east, 30′ south of *A* and at the same elevation as *A*. Point *D* is located 30′ east, 20′ south of *A* and 12′ below *A*. Scale: $1'' = 20'-0''$. Determine the true length, slope, and bearing of the shortest distance between the two lines. *Ans.* T.L. = 17′−6″, Slope = 45°, Bearing = N 72° W

Fig. 3-25

36. A four-sided support frame is shown in Fig. 3-26. A cable must extend from point *X* in a S 60° E direction on a falling slope of 40°. Scale: $\frac{1}{4}'' = 1'-0''$. What is the clearance between the cable and the closest member of the support frame? Show this clearance in all views. *Ans.* Clearance = 2′−9″

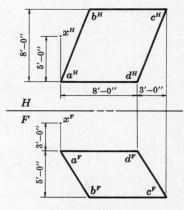

In each of the following problems 37 to 41 determine the true length and bearing of the shortest level distance connecting the skew lines *AB* and *CD*.

Extend the lines if necessary for the solution of the problem. Show the shortest level distance in all views. See Art. 1.7 for the coordinate system of problem layout.

37. $A(4\frac{1}{2}, 1, 5)$ $B(5\frac{1}{2}, 2\frac{1}{2}, 7)$ $C(5, 2, 4\frac{1}{2})$ $D(7, 2\frac{1}{2}, 5\frac{1}{2})$. Scale: $1'' = 1'-0''$.
Ans. T.L. = 1′−9″, Bearing = N 10° W

38. $A(1, 3\frac{1}{2}, 5)$ $B(3, \frac{1}{2}, 4\frac{1}{2})$ $C(1, \frac{1}{2}, 4)$ $D(4, 2, 5\frac{1}{2})$. Scale: $1'' = 40'$.
Ans. T.L. = 5′−6″, Bearing = N 8° W

Fig. 3-26

39. $A(2, 4, 7)$ $B(4\frac{1}{2}, 2, 6)$ $C(3, 2\frac{1}{2}, 5\frac{1}{2})$ $D(5, 4, 5)$. Scale: $1'' = 30'$.
Ans. T.L. = 29′−0″, Bearing = N 14° E

40. $A(3, 3, 6)$ $B(5, 2, 6\frac{1}{2})$ $C(2, 2\frac{1}{2}, 4)$ $D(4, 3\frac{1}{2}, 5\frac{1}{2})$. Scale: $1'' = 50'$.
Ans. T.L. = 55′−6″, Bearing = N 26° W

41. $A(1, 1\frac{1}{2}, 5)$ $B(3, 3, 6)$ $C(3, 1, 6)$ $D(4, 2\frac{1}{2}, 4\frac{1}{2})$. Scale: $1'' = 30'$.
Ans. T.L. = 59′−0″, Bearing = N 68° W

42. Two pipelines are determined by their centerlines *AB* and *CD* shown in Fig. 3-27. Point *B* is located 10′ north, 50′ west of *A* and 35′ above *A*. Point *C* is located 20′ south, 5′ west of *A* and 60′ above *A*. Point *D* is located 40′ north, 30′ west of *A* and 25′ above *A*. Scale: $1'' = 20'-0''$. What would be the true length and bearing of the shortest possible level pipe connecting *AB* and *CD*? Show the level pipe in all views.
Ans. T.L. = 27′−6″, Bearing = N 43°30′ E

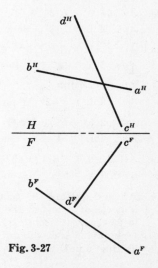

Fig. 3-27

43. Two guy wires, *AB* and *CD*, are located as follows: *B* is 20′ east, 5′ south of *A* and 18′ below *A*. Point *C* is located 12′ due east of *A* and 5′ above *A*. Point *D* is located 20′ east, 10′ north of *A* and 15′ below *A*. Scale: 1″ = 10′−0″. Determine the true length, slope, and bearing of the shortest wire to connect the two guy wires.   *Ans.* T.L. = 10′−0″, Slope = 32°, Bearing = N 47° E

44. Three supporting pipes are shown in Fig. 3-28 below with a cable, *XY*, passing through the pipe frame. Scale: 1″ = 10′−0″. How close does the cable come to the pipe support frame?
       Ans. Distance = 6″

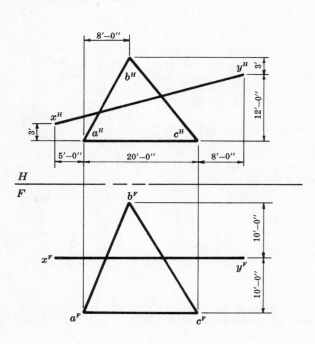

Fig. 3-28

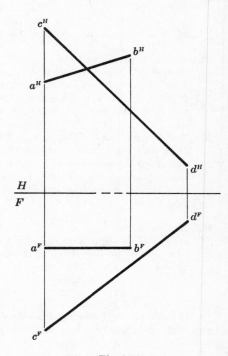

Fig. 3-29

In each of the following problems 45 to 49 determine the true length and bearing of the shortest line of given slope connecting the skew lines *AB* and *CD*.

The given slopes or grades are from line *AB* to line *CD*. Extend the lines if necessary for solution of the problem. Show the shortest line in all views. See Art. 1.7 for the coordinate system of problem layout.

45. $A(1, 2\frac{1}{2}, 6)$ $B(2\frac{1}{2}, 3, 6)$ $C(1, 1, 7)$ $D(3, 3, 5)$. Downward 25%. Scale: 1″ = 1′−0″.
       Ans. T.L. = $10\frac{1}{2}$″, Bearing = S 27° E

46. $A(3, 3, 4\frac{1}{2})$ $B(6, 1, 5)$ $C(5, 2, 5\frac{1}{2})$ $D(6\frac{1}{2}, 3, 4)$. Upward 30°. Scale: 1″ = 20′.
       Ans. T.L. = 13′−9″, Bearing = N 23° E

47. $A(2, 3, 7)$ $B(3, 3\frac{1}{2}, 7)$ $C(\frac{1}{2}, 1, 6\frac{1}{2})$ $D(2\frac{1}{2}, 3, 5\frac{1}{2})$. Downward 30%. Scale: $\frac{1}{4}$″ = 1′−0″.
       Ans. T.L. = 5′−2″, Bearing = S 25° E

48. $A(2\frac{1}{2}, 2, 5)$ $B(5\frac{1}{2}, 1, 4\frac{1}{2})$ $C(5\frac{1}{2}, 2, 6)$ $D(6, 3, 4\frac{1}{2})$. Upward 20°. Scale: 1″ = 30′.
       Ans. T.L. = 54′−6″, Bearing = N 30°30′ E

49. $A(1\frac{1}{2}, 2, 6\frac{1}{2})$ $B(3, 2\frac{1}{2}, 5\frac{1}{2})$ $C(\frac{1}{2}, 1, 6\frac{1}{2})$ $D(2\frac{1}{2}, 3, 5\frac{1}{2})$. Downward 35%. Scale: 1″ = 1′−0″.
       Ans. T.L. = $5\frac{1}{2}$″, Bearing = S 37° W

50. *AB* and *CD* are the centerlines of two natural gas lines. See Fig. 3-29 above. Point *B* is 15′ east, 5′ north of *A* and at the same elevation as *A*. Point *C* is 10′ due north of *A* and 15′ below *A*. Point *D* is 25′ east, 15′ south of *A* and 5′ above *A*. Scale: 1″ = 10′−0″. Determine the shortest connecting pipe having a downward slope of 25%. What would be the true length and bearing of this new connecting pipe? Show the connecting pipe in all views.   *Ans.* T.L. = 10′−6″, Bearing = S 18°30′ E

51. Two pipelines are determined by the lines *AB* and *CD*. Point *B* is 15′ north, 50′ west of *A* and 40′ above *A*. Point *C* is 20′ south, 10′ east of *A* and 60′ above *A*. Point *D* is 45′ north, 30′ west of *A* and 40′ above *A*. Scale: 1″ = 30′–0″. Determine the location and true length of the shortest possible connecting pipe bearing N 20° E and having a rising grade of 35%. Show the connecting pipe in all views.   *Ans.* T.L. = 33′–6″

52. The flight direction of a jet plane at *A* is N 45° E as shown in Fig. 3-30 below. It is gaining altitude at the rate of 300′ in 1000′. Scale: 1″ = 1000′. Determine the clearance between the flight path and an obstruction represented by line *XY*. $A(4\frac{1}{4}, 2\frac{1}{2}, 6)$ $X(5\frac{1}{4}, 2, 6\frac{1}{4})$ $Y(7, 3\frac{1}{2}, 7\frac{1}{2})$.

Ans. Clearance = 765′

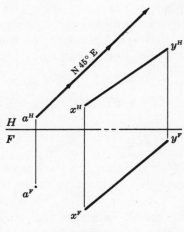

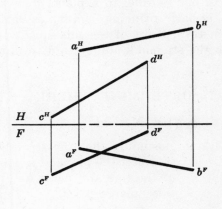

Fig. 3-30   **Fig. 3-31**

53. The centerlines of two mine shafts are determined by the lines *AB* and *CD*. See Fig. 3-31 above. Point *B* is 100′ east, 20′ north of *A* and 20′ below *A*. Point *C* is 60′ south, 25′ west of *A* and 25′ below *A*. Point *D* is 60′ east, 10′ south of *A* and 15′ above *A*. A new connecting tunnel, *XY*, is driven between the two main shafts. Its bearing is N 45° W and slopes downward 30° to the northwest. Scale: 1″ = 50′. Determine the true length of the connecting tunnel. Show tunnel *XY* in all views.
Ans. T.L. = 34′–6″

54. From point *A*, at an elevation of 1800′, a tunnel bears S 20° E on a downward grade of 35%. Another tunnel starts at *B* which is located 130′ south, 50′ west of *A* and 120′ above *A*. This second tunnel bears S 80° E on a downward grade of 20%. Scale: 1″ = 100′. What would be the true length, slope, and bearing of the shortest connecting tunnel? What would be the elevation at both ends of the connecting tunnel? Show the proposed connecting tunnel in all views. Use the line method. Let point *X* represent the lower end of the connecting tunnel.
Ans. T.L. = 145′, Slope = 71°, Bearing = N 23° W, Elev. *X* = 1760′, Elev. *Y* = 1895′

55. *AB* and *CD* determine the centerlines of two mine tunnels. Point *B* is 350′ south, 180′ west of *A* and 150′ above *A*. Point *C* is 350′ south, 300′ west of *A* and 50′ above *A*. Point *D* is located 200′ south, 50′ east of *A* and 100′ above *A*. Scale: 1″ = 200′. What would be the true length and bearing of the shortest possible level connecting tunnel between the two given mine tunnels? Show the level tunnel in all views.   *Ans.* T.L. = 70′, Bearing = N 33° W

Chapter 4

Piercing Points and Plane Intersections

It should be obvious to the student of Descriptive Geometry that a line, which neither lies in nor parallel to a plane, must intersect the plane. The knowledge and application of this basic principle is very essential to the solution of problems involving Descriptive Geometry. Many of the subsequent problems in this text are dependent for their solutions upon the student's ability to determine the point where a line intersects a plane.

4.1 To DETERMINE WHERE a LINE INTERSECTS a PLANE

A. Edge-View Method

Analysis: Assuming that the straight line is neither parallel to nor in the plane, it will intersect the plane at a point common to both the line and the plane. The limits of the line and plane as given may have to be extended in order to determine this "pierce point".

Since an edge-view of the plane contains all points in the plane, the view showing the plane as an edge will also show the point where the line pierces the plane.

Note: Another method would involve showing the line as a point. This point view of the line will contain the pierce point common to both the line and the plane.

Example 1: (Given Plane Appears as an Edge) In Fig. 4-1 below, the plane *ABCD* and the line *XY* are given in both the plan and front elevation views. The plane appears as an edge in the plan view. The intersection of the line and the plane is determined by point *P* which is common to both. Using careful visualization, we notice that the *YP* portion of the line lies wholly in front of the plane and therefore will appear visible in the front view. The plan view shows that the *XP* portion of the line lies wholly behind the plane and, therefore, will be hidden in the front view.

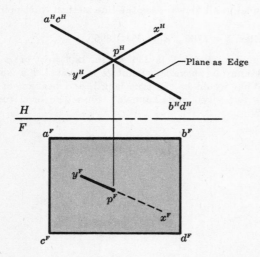

Fig. 4-1. A Line Intersecting a Vertical Plane

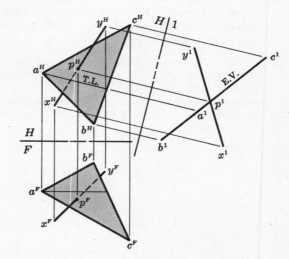

Fig. 4-2. A Line Intersecting an Oblique Plane

Example 2: (Oblique Plane) In Fig. 4-2 above, the plane ABC and the line XY are given in both the plan and front elevation views. Draw an auxiliary elevation view showing the plane as an edge. The pierce point P in this view is the point of intersection which can now be projected back to the plan and front elevation views. Careful visualization reveals the visible and hidden portions of the line.

Note: The pierce point could also have been determined by showing the plane as an edge in an inclined view projected from the front view.

B. Two-View Cutting Plane Method

Analysis: The intersection of a line and an oblique plane can be determined by using a vertical cutting plane which contains the given line. The line of intersection of the cutting plane with the oblique plane and the given line must intersect or be parallel because they both lie in the vertical cutting plane. (See Fig. 4-3.) Since the cutting plane appears as an edge in the plan view, the relationship between the line of intersection and the given line is not apparent in the plan view. The related view, however, reveals this relationship. Should the two lines intersect in the related view, it is evident that the point of intersection is common to both the given plane and the given line and therefore determines the pierce point of the given line and given plane.

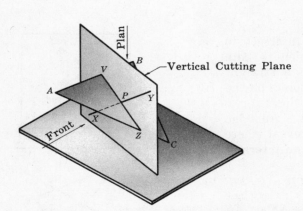

Fig. 4-3. A Pictorial View of the Vertical Cutting Plane Method

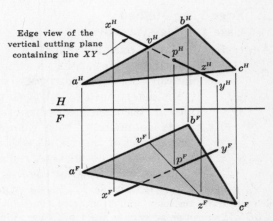

Fig. 4-4. A Line Intersecting an Oblique Plane (Vertical Cutting Plane Method)

Example: In Fig. 4-4 the oblique plane ABC and the line XY are given in both the plan and front elevation views. A vertical cutting plane, coincidental with and containing the given line XY, appears as an edge in the plan view. The intersection of the given plane ABC and the vertical cutting plane containing XY is the line VZ. The lines XY and VZ both lie in the vertical cutting plane and intersect each other at point P in the front view. Since point P is on line VZ, it is also on plane ABC because line VZ is on plane ABC. Therefore point P is the required point, being common to both the given line XY and the given plane ABC. It can now be projected to the related view. Use careful visualization to determine what portion of the line should be visible in each view.

Note: If line VZ had appeared parallel to XY in the front view, it would have indicated that the line XY was parallel to plane ABC and therefore it would have no point of intersection with the given plane.

Another similar solution would be that of having the cutting plane shown as an edge in the front view and containing the given line (see Fig. 4-5).

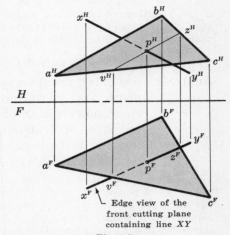

Fig. 4-5. A Line Intersecting an Oblique Plane (Front Cutting Plane Method)

Even though the two-view method has the advantage of quick construction and minimum space requirements, the edge-view method is usually the easier method for the beginning student to understand.

4.2 INTERSECTION of PLANES

One of the most common problems encountered in Descriptive Geometry is that of determining the line of intersection of two planes. As mentioned previously, a plane is considered to be indefinite in extent, and, likewise, the line of intersection between two planes is also considered to be of an indefinite length. On the other hand, if a particular plane surface is "closed", or limited, the line of intersection between that plane and another plane would then be considered limited.

The intersection of any two oblique planes is a straight line which is common to both planes. Any two points common to both planes will, therefore, determine the position of the line of intersection.

The three general methods used in determining the line of intersection between any two planes are the "edge-view" method, the "two-view pierce point" method, and the "two-view cutting plane" method.

A. Edge-View Method
Case I: Edge-view Given

Analysis: Two non-parallel planes will intersect in a straight line which is common to both planes. Since the direction of a straight line is determined by any two points on the line, it becomes necessary to locate two points common to both planes. An edge view of one of the planes will reveal where any two lines on the other plane intersect the plane shown as an edge.

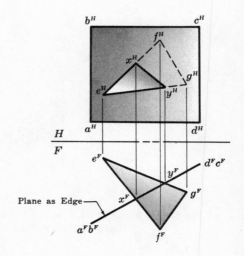

Example: In Fig. 4-6 the two planes, *ABCD* and *EFG*, are given in the plan and front elevation views. The plane *ABCD* is shown as an edge in the front view. Lines *EF* and *EG* on plane *EFG* intersect the plane *ABCD* at points *X* and *Y* respectively. Since points *X* and *Y* are common to both planes, they must, therefore, lie on the line of intersection. To locate the line of intersection on the plan view, simply project points *X* and *Y* to their corresponding positions. The front view shows the *EXY* portion of plane *EFG* to lie wholly above plane *ABCD*; therefore *EXY* will be visible in the plan view.

Fig. 4-6.

Case II: Edge-view Not Given

Analysis: Assuming that two views of each plane are given, an additional view showing both planes, one as an edge, will show two points common to both planes, thus determining the line of intersection.

Example: In Fig. 4-7 the plan and front elevation views of planes *ABC* and *DEFG* are given. An auxiliary elevation view is drawn showing both planes, one of which, *ABC*, appears as an edge. In this view, line *EG* of plane *DEFG* pierces *ABC* at point *X*, and line *DF* pierces plane *ABC* at *Y*. Points *X* and *Y* are two points on the common line of intersection between the two planes.

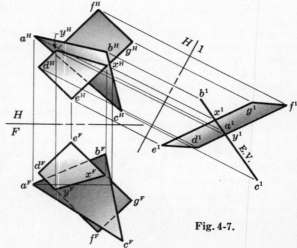

Fig. 4-7.

In this problem the two given planes are limited and the line of intersection may be regarded as terminating at the points X and Y. Careful visualization will determine the visibility of the other lines in the planes.

B. Two-View Pierce Point Method

Analysis: Following the method of Art. 4.1-*B*, a line on one of the planes may be selected and the point of its piercing the other plane will be one point common to both planes. Repeat the procedure using another line and determining its pierce point with the other plane. These two "pierce points" determine the line of intersection of the two planes.

Example: The two oblique planes *ABC* and *DEFG* are given in both the plan and front elevation views in Fig. 4-8 below. Using the method of Art. 4.1-*B*, the line *DF* of plane *DEFG* is found to pierce *ABC* at point *Y*. Likewise, the line *EG* of plane *DEFG* pierces the plane *ABC* at point *X*. Points *X* and *Y*, being common to both planes, determine the line of intersection of both planes.

Careful visualization will determine the visibility of the other lines in the plane.

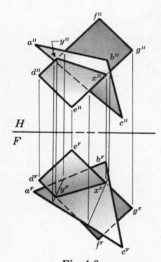

Fig. 4-8.
Intersection of Two Oblique Planes
(Two-view Pierce Point Method)

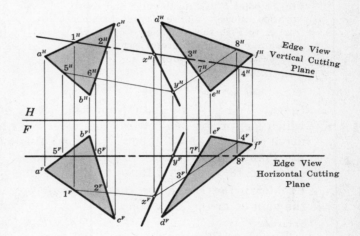

Fig. 4-9.   Intersection of Two Oblique Planes
(Two-view Cutting Plane Method)

C. Two-View Cutting Plane Method

Analysis: If two given oblique planes are intersected by a third plane which appears as an edge in one view, the three planes will meet at a common point, unless the third plane should happen to be parallel to the intersection of the two given planes. The method used to determine the location of two or more points on the line of intersection is that which is explained in the article dealing with the edge-view method.

Example: The two oblique planes *ABC* and *DEF* are given in both the plan and front elevation views in Fig. 4-9 above. A vertical cutting plane intersects the two given planes at points 1-2 and 3-4. The lines from 1 to 2 and from 3 to 4 are extended in the front view to locate point *X* which is a point common to both planes extended and is, therefore, on the line of intersection of the two planes. Point *X* is located on the plan view by projecting it up to the edge view of the cutting plane.

To locate a second point on the line of intersection, a horizontal cutting plane intersects the given planes at points 5-6 and 7-8. The lines from 5 to 6 and from 7 to 8 are extended in the plan view to locate point *Y* which is a point common to both planes extended and is, likewise, on the line of intersection of the two planes. To locate point *Y* in the front view, simply project down to the edge view of the horizontal cutting plane.

Solved Problems

1. **Given:** Plane ABC and line MN in Fig. 4-10. $A(1,1,4)$ $B(4,1,5\frac{1}{2})$ $C(3,2\frac{1}{2},4)$ $M(1\frac{3}{4},2\frac{1}{4},5\frac{1}{4})$ $N(3\frac{3}{4},\frac{3}{4},3\frac{3}{4})$. See Art. 1.7 for the coordinate system of problem layout. Scale: $\frac{1}{4}'' = 1'-0''$.

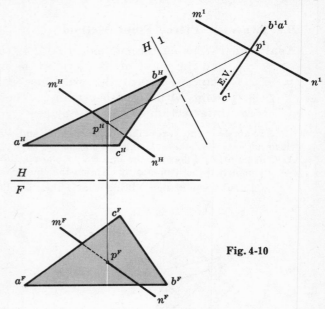

Problem: Using the edge-view method, determine the intersection of line MN with plane ABC. What is the map distance from point A to the pierce point? Show proper visibility.

Solution:

Using the given data, draw the plan and front elevation views of line MN and plane ABC. Draw auxiliary elevation view 1 to show plane ABC as an edge with line MN intersecting the plane at point P. Project point P back to the plan and front elevation views. Measure the distance from A to P in the plan view for the required map distance. Careful visualization will reveal proper visibility.

Ans. Distance $= 3'-7\frac{3}{4}''$

Fig. 4-10

2. **Given:** Points A, B, and C represent a vein of ore. Point Y is the portal of a tunnel which has progressed toward the vein as far as point X. Point B is 24' east, 18'-6'' north of A and 15'-6'' above A. Point C is 41'-6'' east, 13' south of A and 21' below A. Point X is 46' east, 7' south of A and 12' above A. Point Y is 61' east, 13' south of A and 18' above A. Refer to Fig. 4-11 below. Scale: $1'' = 40'$.

Problem: Determine how much further the tunnel must be extended in order to reach the vein of ore.

Solution:

Using the given data, draw the plan and front elevation views of plane ABC and line XY. Show the plane ABC as an edge in an elevation view, projecting line XY into the same view. Extend line XY until it intersects the edge view of the plane. Project the entire tunnel back to the plan view. Now having two views of the line from X to the plane, an auxiliary elevation view can be drawn to determine the true length of that distance. (See Art. 4.1).   *Ans.* Distance $= 32'$

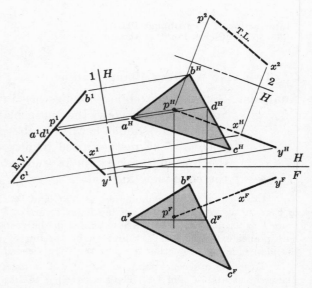

Fig. 4-11

3.   Given: Planes *ABC* and *EFG* shown in Fig. 4-12 below.   Point *B* is 2″ east, 3″ north of *A* and 3″ above *A*.   Point *C* is 2½″ east, 1″ north of *A* and 2″ above *A*.   Point *E* is located ½″ due north of *A* and 2″ above *A*.   Point *F* is 2½″ due north of *A* and 3″ above *A*.   Point *G* is 2½″ east, 1½″ north of *A* and 1″ above *A*.   Scale: ¾″ = 1″.

Problem: Using the edge-view method, determine the bearing of the line of intersection between the two planes.

Solution:

        Using the given data, draw the plan and front elevation views of planes *ABC* and *EFG*. An auxiliary elevation view is drawn showing both planes, one of which, *ABC*, appears as an edge. Auxiliary elevation view 1 locates two points common to both planes. Label these two points *X* and *Y*. Project points *X* and *Y* to the plan and front views and connect them to locate the line of intersection in these two views. Measure the bearing of the line of intersection in the plan view. Careful visualization will determine proper visibility.   *Ans.* Bearing = N 6° W

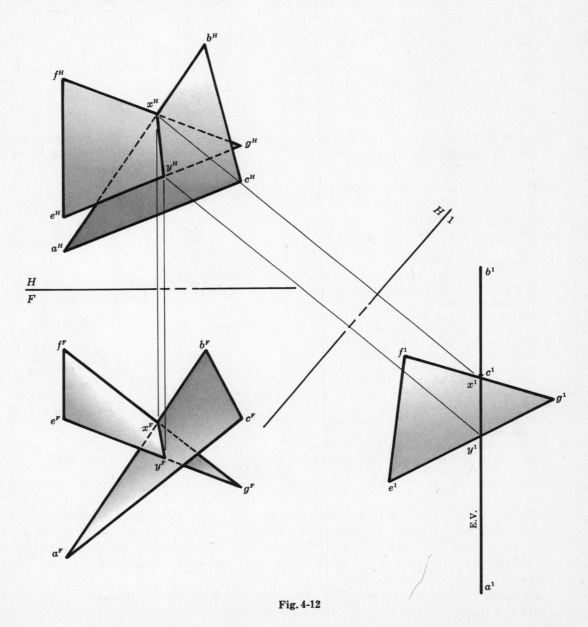

Fig. 4-12

4. **Given:** Fig. 4-13 shows the plane of a roof section and the top of an antenna, point X. Scale: $\frac{1}{16}'' = 1'-0''$.

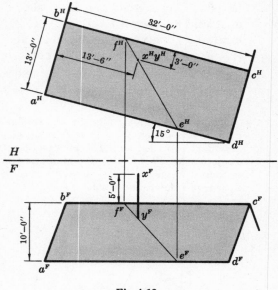

Problem: Using the cutting plane method, determine the true length of the vertical antenna. Show the antenna in the front elevation view.

Solution:

Using the cutting plane method of Art. 4.1-B, draw a line through the plan view of point X. This line represents the plan view of a vertical cutting plane and also its intersection with the plane of the roof section. Label the intersections of this cutting plane with AD and BC, points E and F respectively. Project points E and F to their corresponding positions in the front view. Connect E and F in the front view. From point X in the front view, draw a vertical line down till it meets the line EF. Label this intersection point, Y. The line XY in the front view is the true length of the antenna.

Ans. T.L. $= 7'-6''$

Fig. 4-13

5. **Given:** Planes ABC and $DEFG$ shown in Fig. 4-14. $A(\frac{3}{4}, 2\frac{1}{2}, 5\frac{1}{4})$ $B(3\frac{1}{2}, 3, 6\frac{3}{4})$ $C(2\frac{3}{4}, \frac{3}{4}, 4\frac{3}{4})$ $D(2, 3\frac{1}{4}, 5\frac{3}{4})$ $E(3, 3\frac{1}{4}, 5)$ $F(2, 1, 6\frac{1}{4})$ $G(1, 1, 7)$. See Art. 1.7 for the coordinate system of problem layout.

Problem: Using the edge-view method, show the intersection of the two planes in all views. Determine the bearing of the line of intersection.

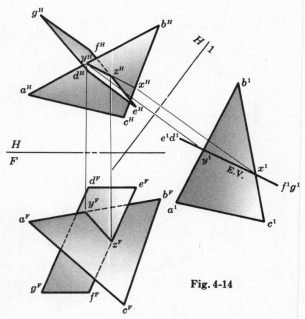

Solution:

Using the given data, locate the plan and front elevation views of both planes. Draw an auxiliary elevation view showing both planes, one of which, $DEFG$, appears as an edge. Auxiliary elevation view 1 locates two points, X and Y, which lie on the line of intersection. Project X and Y back to the plan view to determine the bearing of the line of intersection. You will notice that the line of intersection in the plan view extends from point Y to the intersection with line EF. Label this intersection point, Z. Project Y and Z to their corresponding positions in the front view. Use careful visualization to determine proper visibility.

Ans. Bearing $=$ N $64°$ W

Fig. 4-14

6. **Given:** Plane $ABCD$ is intersected by a control cable, XY. Point B is located 100' due east of A and 50' below A. Point C is located 75' due south of A and 65' below A. Point D is 100' east, 75' south of A and 115' below A. Point X is 20' west, 25' south of A and 120' below A. Point Y is located 120' east, 25' south of A and 25' above A. See Fig. 4-15 below. Scale: $1'' = 100'$.

Problem: Locate the pierce point of the control cable with the plane by using the edge-view method. Determine the true length and bearing of a line from point A to the pierce point. The pierce point is how much lower than A?

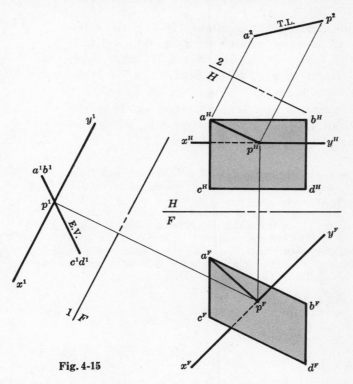

Fig. 4-15

Solution:

Draw the plan and front elevation views of line *XY* and plane *ABCD*. Draw an inclined view showing the plane as an edge and the line *XY* piercing it at point *P*. Project point *P* back to the front elevation and plan views. Draw a line from *A* to *P* in the plan view. Measure the bearing of line *AP* in this view. Draw auxiliary elevation view 2 in order to obtain the true length of *AP*. Measure the elevational relationship between *A* and *P* in either the front view or in auxiliary elevation view 2. Show the proper visibility of the line and plane by careful visualization.

Ans. T.L. = 73′–6″, Bearing = S 63°30′ E, Elevation of P = 40′ lower than A

7. **Given:** A mine tunnel, in the direction *XY* as shown in Fig. 4-16, is driven toward a vein of coal represented by plane *ABC*. Point *B* is 80′ east, 80′ north of *A* and 80′ below *A*. Point *C* is located 120′ due east of *A* and 40′ above *A*. Point *X* is 10′ west, 50′ north of *A* and 50′ above *A*. Point *Y* is 40′ east, 15′ north of *A* and 25′ above *A*. Scale: 1″ = 100′.

Problem: How long will it take the miners to reach the vein of coal, starting from point *X*, if they average 12′ per hour? What is the total length of the tunnel necessary to reach the vein?

Solution:

Using the given data, draw the plan and front elevation views of plane *ABC* and tunnel *XY*. As explained in Art. 4.1-*B*, the cutting plane method is used to determine points *D* and *E* in both views. The intersection of line *DE* in the front view with the line *XY* extended will locate the pierce point *P*. Project point *P* up to the plan view. Draw an inclined view 1 to find the true length of the entire tunnel. Divide the true length of *XP* by 12 to determine the total time required to reach the vein.

Ans. T.L. = 89′, Time = 7 hrs. 25 mins.

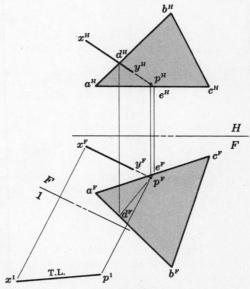

Fig. 4-16

8. **Given:** A square base right pyramid $ABCD$ with vertex O and a line XY passing through the pyramid. $A(3,3,7)$ $B(4,3,8)$ $C(5,3,7)$ $D(4,3,6)$ $O(4,5,7)$ $X(3,3\frac{1}{2},8)$ $Y(5\frac{1}{2},3\frac{1}{2},6)$.   Refer to Fig. 4-17.   See Art. 1.7 for the coordinate system of problem layout.   Scale: $\frac{1}{2}'' = 1''$.

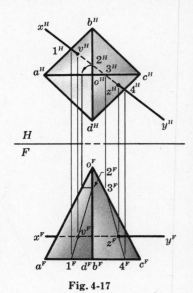

Problem: Determine the true distance between the pierce points of line XY and the pyramid.   Use two views only and show proper visibility.

Solution:

Using the given coordinate data, draw the plan and front elevation views of the pyramid and line.   By means of the cutting plane method of Art. 4.1-*B*, the pierce points V and Z are determined in the front view.   Project points V and Z from the front view up to the line XY in the plan view.   Pierce points V and Z are thus located in the plan view.   Since the line XY is a level line, the distance between points V and Z in the plan view is a true length distance.   Careful visualization will determine proper visibility.   *Ans.* T.L. $= 1\frac{1}{16}''$

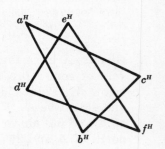

Fig. 4-17

Supplementary Problems

9. In each of the following cases determine the point of intersection between the given plane ABC and the line XY.   Use the edge-view method.   Show proper visibility.   If the scale of $1'' = 10'$ is used, determine the map distance from point A to the pierce point.   See Art. 1.7 for the coordinate system of problem layout.

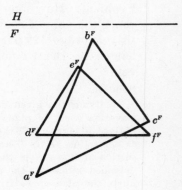

(*a*) $A(1,2\frac{1}{2},4)$ $B(1\frac{1}{2},1,6)$ $C(3\frac{1}{2},1\frac{1}{2},5)$ $X(1,1\frac{1}{2},6)$ $Y(3\frac{1}{2},1,4)$
Ans. $16'-4''$

(*b*) $A(1\frac{1}{2},1\frac{1}{2},5)$ $B(2\frac{1}{2},3,6\frac{1}{2})$ $C(4,1,4\frac{1}{2})$ $X(3\frac{1}{2},1,6)$
$Y(2\frac{1}{2},2\frac{1}{2},4\frac{1}{2})$ *Ans.* $15'-6''$

(*c*) $A(1,1,6)$ $B(2\frac{1}{2},2\frac{1}{2},5)$ $C(3\frac{1}{2},\frac{1}{2},6\frac{1}{2})$ $X(2,\frac{1}{2},5)$ $Y(3\frac{1}{2},2,7)$
Ans. $17'-0''$

(*d*) $A(1\frac{1}{2},3,7)$ $B(2\frac{1}{2},1,5)$ $C(3\frac{1}{2},1\frac{1}{2},6\frac{1}{2})$ $X(1\frac{1}{2},1,5\frac{1}{2})$
$Y(3,2\frac{1}{2},7)$ *Ans.* $11'-10''$

(*e*) $A(1,3,5\frac{1}{2})$ $B(3,3,5)$ $C(4,1\frac{1}{2},6\frac{1}{2})$ $X(2,2,5)$ $Y(3\frac{1}{2},3,5\frac{1}{2})$
Ans. $20'-7''$

10. Line XY and plane ABC are determined as follows: $A(1,1,4)$ $B(4,1,5\frac{1}{2})$ $C(3,2\frac{1}{2},4)$ $X(1\frac{3}{4},2\frac{1}{4},5\frac{1}{4})$ $Y(3\frac{3}{4},\frac{3}{4},3\frac{3}{4})$. See Art. 1.7 for the coordinate system of problem layout.   Using the two-view cutting plane method, find the approximate coordinate location of the pierce point of line XY with plane ABC.
Ans. Pierce point $= (2\frac{3}{4},1\frac{1}{2},4\frac{1}{2})$

11. Two planes, ABC and DEF, are shown in Fig. 4-18. $A(1,1,6\frac{1}{2})$ $B(2,3\frac{1}{2},4\frac{1}{2})$ $C(3,2,5\frac{1}{2})$ $D(1,1\frac{3}{4},5\frac{1}{4})$ $E(1\frac{1}{4},3,6\frac{1}{2})$ $F(3,1\frac{3}{4},4\frac{1}{2})$.
Determine the intersection of the two planes.   Show proper visibility.   What is the bearing of the line of intersection?
Ans. Bearing $= N 77° W$

Fig. 4-18

12. A bent plate, *ABCD*, is located as follows: Point *B* is 10′ east, 15′ north of *A* and 25′ below *A*. Point *C* is 40′ east, 10′ north of *A* and 15′ below *A*. Point *D* is 35′ east, 10′ south of *A* and 5′ above *A*. The "bend line" is *AC*. See Fig. 4-19 below. A cable, *XY*, must pass through the bent plate. Point *X* is 10′ north, 8′ west of *A* and 10′ below *A*. Point *Y* is 40′ east, 5′ south of *A* and 5′ below *A*. Scale: 1″ = 10′. Using two views only, determine the point, or points, where the cable will pass through the bent plate. What is the elevational relationship between the pierce point, or points, and *A*?

Ans. Pierce point is 8′–6″ lower than *A*.

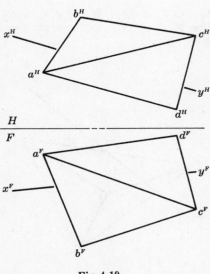

Fig. 4-19

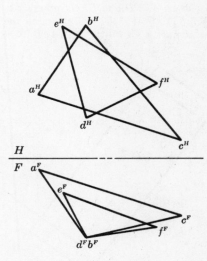

Fig. 4-20

13. The two planes *ABC* and *DEF* shown in Fig. 4-20 above intersect each other. Point *B* is 6′ north, 4′ east of *A* and 6′ below *A*. Point *C* is 4′ south, 12′ east of *A* and 4′ below *A*. Point *D* is 2′ south, 4′ east of *A* and 6′ below *A*. Point *E* is 6′ north, 2′ east of *A* and 2′ below *A*. Point *F* is 1′ north, 10′ east of *A* and 5′ below *A*. Scale: ¼″ = 1′–0″. Using the edge-view method, show the line of intersection between the two planes. Determine the visibility in both plan and front views. What is the bearing of the line of intersection?   *Ans.* Bearing = N 68° W

14. The plane of *ABC* is intersected by a level line *XY*. Point *B* is 10′ east, 20′ north of *A* and 15′ below *A*. Point *C* is 30′ east, 5′ north of *A* and 10′ above *A*. The line *XY* is at an elevation of 3′ below *A* with point *X* located 2′ due north of *A* and point *Y* located 3′ due south of *C*. Scale: 1″ = 10′. Determine the pierce point of line *XY* with the plane of *ABC*. What is the bearing of a line drawn from the pierce point to *A*?

Ans. Bearing = S 45° E

15. Two planes, *ABC* and *DEF*, intersect each other. Refer to Fig. 4-21. Point *B* is 20′ east, 10′ north of *A* and 25′ above *A*. Point *C* is 30′ east, 5′ south of *A* and 10′ above *A*. Point *D* is 12′ east, 10′ north of *A* and 20′ above *A*. Point *E* is 25′ east, 10′ south of *A* and 5′ above *A*. Point *F* is 35′ east, 13′ north of *A* and 10′ above *A*. Scale: 1″ = 10′. Locate the line of intersection between the two given planes and show this line in both the plan and front views. Determine the visibility of the planes in all views. What is the bearing of the line of intersection? Use the edge-view method.

Ans. Bearing = N 42° W

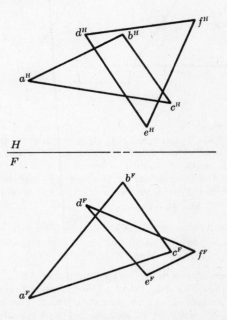

Fig. 4-21

16. A rectangular frame, *ABCD*, is supported by two wooden braces, *EF* and *GH*. Determine the points where the braces pass through the plane *ABCD*. How much of the braces must be removed if they are not to extend beyond the plane of the frame? See Fig. 4-22 below. Point *B* is 4′ east, 3′ north of *A* and 8′ above *A*. Point *C* is 13′ east, 9′ south of *A* and 8′ above *A*. Point *D* is 9′ east, 12′ south of *A* and at the same elevation as *A*. Point *E* is 1′–6″ east, 6″ south of *A* and 8′–9″ above *A*. Point *F* is 7′–6″ east, 4′ north of *A* and at the same elevation as *A*. Point *G* is 9′ east, 10′–6″ south of *A* and 8′–9″ above *A*. Point *H* is 15′ east, 6′ south of *A* and at the same elevation as *A*. Scale: $\frac{1}{4}$″ = 1′–0″.
Ans. Remove 3′–10″

17. Two planes, *ABCD* and *EFG*, intersect each other. $A(2, 2, 5)$ $B(2, 2, 8)$ $C(5, 4, 8)$ $D(5, 4, 5)$ $E(2\frac{1}{2}, 3\frac{1}{2}, 6\frac{1}{2})$ $F(3\frac{1}{2}, 2, 7\frac{1}{2})$ $G(4\frac{1}{2}, 3, 5\frac{1}{2})$. Determine the bearing of the line of intersection between the two planes. Show complete visibility.   *Ans.* Bearing = N 30°30′ W

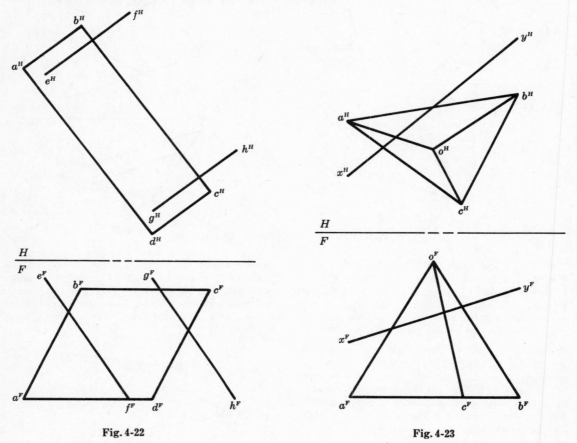

Fig. 4-22   Fig. 4-23

18. A pyramid with base *ABC* and vertex *O* is shown in Fig. 4-23 above. A line *XY* passes through the pyramid. Point *B* is 12′ east, 2′ north of *A* and at the same elevation as *A*. Point *C* is 8′ east, 6′ south of *A* and at the same elevation as *A*. Point *O* is 6′ east, 2′ south of *A* and 10′ above *A*. Point *X* is 4′ due south of *A* and 4′ above *A*. Point *Y* is 12′ east, 6′ north of *A* and 8′ above *A*. Scale: $\frac{1}{4}$″ = 1′–0″. Show the intersection of the line *XY* with the pyramid in both the plan and front views. If point *X* is located 2′ further south of *A*, what would be the distance between the piercing points of the new line *XY* and the pyramid?   *Ans.* T.L. = 1′–6″

19. Plane *ABC* and line *XY* are located as follows: $A(1, 1\frac{1}{4}, 4)$ $B(2\frac{1}{4}, 2\frac{3}{4}, 5\frac{3}{4})$ $C(4, 1, 4)$ $X(1, 1\frac{1}{4}, 5\frac{1}{2})$ $Y(2\frac{1}{2}, 2\frac{1}{4}, 3\frac{1}{2})$. Determine the approximate coordinate location for the intersection of line *XY* with plane *ABC*.   *Ans.* Pierce point = $(1\frac{3}{4}, 1\frac{3}{4}, 4\frac{1}{2})$

20. A vein of ore is determined by plane *ABC*. A rising mine tunnel, *XY*, is driven toward the vein of ore. See Fig. 4-24 below. Point *B* is 100′ east, 75′ south of *A* and 85′ above *A*. Point *C* is 200′ east, 50′ north of *A* and 25′ above *A*. Point *X* is 150′ east, 50′ south of *A* and 25′ below *A*. Point *Y* is 250′ east, 90′ south of *A* and 75′ below *A*. Scale: 1″ = 50′. How much must the tunnel be extended in order to reach the vein of ore?   *Ans.* 116′

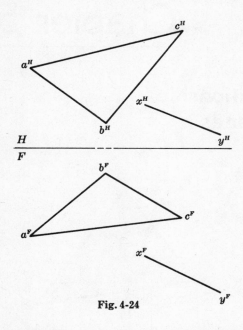

Fig. 4-24

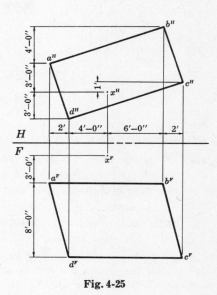

Fig. 4-25

21. The horizontal base of a pyramid is represented by an equilateral triangle, 3″ on each side. The three lateral faces of the pyramid makes angles of 45°, 60°, and 75° with the base plane. Construct a plan view and the necessary elevation views of the pyramid to determine its altitude. Scale: 12″ = 1′−0″. *Ans.* Altitude = $1\frac{7}{16}''$

22. Two planes, *ABCD* and *EFG*, intersect each other. $A(2, 4, 6)$ $B(2, 2, 6)$ $C(5, 4, 7\frac{1}{2})$ $D(5, 2, 7\frac{1}{2})$ $E(2\frac{1}{2}, 3\frac{1}{2}, 7\frac{1}{2})$ $F(3\frac{1}{2}, 2\frac{1}{2}, 6)$ $G(5, 3, 6\frac{1}{2})$. See Art. 1.7 for the coordinate system of problem layout. Determine the bearing of the line of intersection. Show complete visibility.   *Ans.* Bearing = N 64° E

23. Fig. 4-25 above shows the plan and front elevation views of a roof section *ABCD*. Point *X* represents the top of a television antenna. Using two views only, determine the true length of the vertical antenna. Scale: $\frac{1}{4}'' = 1'-0''$.   *Ans.* T.L. = 9′−0″

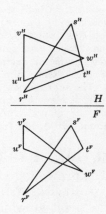

24. Planes *RST* and *UVW* are given in the adjacent Fig. 4-26. Point *S* is 2″ east, 3″ north of *R* and 3″ above *R*. Point *T* is $2\frac{1}{2}''$ east, 1″ north of *R* and 2″ above *R*. Point *U* is located $\frac{1}{2}''$ due north of *R* and 2″ above *R*. Point *V* is $2\frac{1}{2}''$ due north of *R* and 3″ above *R*. Point *W* is $2\frac{1}{2}''$ east, $1\frac{1}{2}''$ north of *R* and 1″ above *R*. Scale: 12″ = 1′−0″. Determine the line of intersection between the two planes by using the two-view cutting plane method. What is the bearing of the line of intersection?   *Ans.* Bearing = N 6° W

Fig. 4-26

Chapter 5

Perpendicular Relationships:
Lines and Planes

5.1 PROJECTION of a POINT on a PLANE

Analysis: A perpendicular line from the given point to the given plane will determine the projection of the point on the plane. It is therefore necessary to locate the intersection of the perpendicular line with the given plane.

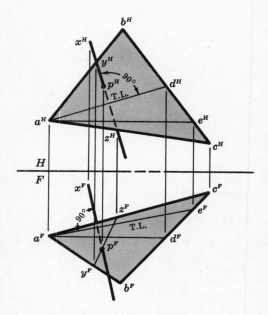

Fig. 5-1.
Projection of a Point on a Plane

Example: In Fig. 5-1, the plane *ABC* and the point *X* are given in both the plan and front elevation views. The problem requires locating the projection of point *X* on plane *ABC*. The first step is to find the perpendicular from the point to the plane. Draw true length lines *AD* (horizontal) and *AE* (frontal) in both the plan and front elevation views respectively. Lines *AD* and *AE* are derived from the geometric theorem stating that a line is perpendicular to a plane if it is perpendicular to two non-parallel lines in the plane. From point *X* in both views, draw a line of indefinite length perpendicular to the true length lines. The second step is to locate the intersection of this perpendicular line with the given plane. Assume a vertical cutting plane through the perpendicular from *X* in the plan view. The intersection of the cutting plane and given plane *ABC* is along line *YZ*. The intersection of *YZ* and the perpendicular from *X* in the front elevation view will locate the projection of the point on the plane. The plan view of this point is obtained by simple projection from the front elevation view.

5.2 PROJECTION of a LINE on a PLANE

Analysis: The projection of a line on a plane is based upon the same principle as shown in the previous article. It was shown that the projection of a point on a plane is the point at which a perpendicular from the point to the plane intersected the plane. The projection of a line on a plane is determined simply by projecting any two points of that line on the given plane. Once the two point projections are determined, a line is drawn through them to obtain the required projections of the line on the plane.

Example 1: (Two-View Cutting Plane Method) In Fig. 5-2 below, the plane *ABC* and the line *XY* are given in both the plan and front elevation views. Draw true length lines *AD* and *CE* in both plan and front views, respectively. From points *X* and *Y* in each view, draw perpendiculars to the true length lines. Assume vertical cutting planes through the perpendiculars from *X* and *Y* in the plan view. The intersections of the cutting planes with the given plane are along lines *ST* and *UV*. The intersection of *ST* and the perpendicular from *X* in the front view will locate the projection of point *X* in the front view. Label this point *P*. The intersection of *UV* and the perpendicular from *Y* in the front view will locate the projection of point *Y* in the front view. Label this point *Q*. With the projection of both *X* and *Y* determined, simply connect *P* and *Q* in the front view to determine the required line. The plan view of points *P* and *Q* is obtained by simply projecting from the front view.

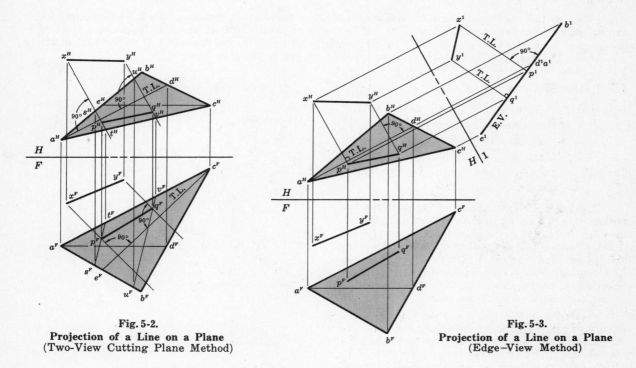

Fig. 5-2.
Projection of a Line on a Plane
(Two-View Cutting Plane Method)

Fig. 5-3.
Projection of a Line on a Plane
(Edge–View Method)

Example 2: (Edge-View Method) In Fig. 5-3 above, the plane ABC and the line XY are given in both the plan and front elevation views. An auxiliary elevation view is drawn showing the plane as an edge as well as the line XY. Draw perpendicular projection lines from each end of the line XY to the edge view of the plane. Label the intersection of these projections from X and Y as P and Q, respectively. The auxiliary elevation view 1 shows the perpendicular projection lines in their true length, so they must appear in the plan view parallel to the H-1 folding line. Project P and Q back to the plan view until they intersect the projections from X and Y which are parallel to the H-1 folding line. These intersections will determine the plan view of the given line projected on the plane ABC. Simple projection and transfer of distances will locate the front elevation view of the given line projected on the plane.

5.3 SHORTEST DISTANCE from a POINT to a LINE

A. Line Method

Analysis: The shortest distance from a given point to a given line is the perpendicular distance. When the given line appears in its true length, the shortest distance will appear as a perpendicular to it. If a view is drawn showing the given line as a point, the required distance will appear in its true length.

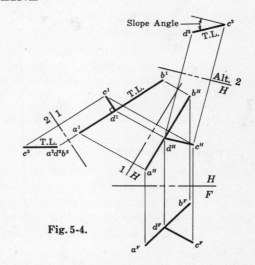

Fig. 5-4.

Example: Fig. 5-4 shows the given line AB and the point C in both the plan and front elevation views. To determine the shortest distance between point C and the line AB, we first show the line AB in its true length in either an auxiliary elevation view as shown, or by taking an inclined view off the front view. The shortest distance will appear in this view as the perpendicular from point C to the line AB. Label this point of intersection, D.

To obtain the true length of the shortest distance, CD, draw an inclined view which shows the line AB as a point. Since folding line 1-2 is parallel to CD in view 1, view 2 will show the shortest distance in its true length. Project point D back to the other views.

Note: If the problem also involves finding the slope of the shortest distance, view 2 would be omitted and an alternate auxiliary elevation view would be drawn to determine both the true length and slope of the shortest distance. This procedure, of course, would be preceded by locating point *D* on line *AB* in the plan view. The bearing of the shortest distance would appear in the plan view.

B. Plane Method

Analysis: A line and a given point not on the line determine a plane. If the plane is shown in its true size, then a perpendicular from the point to the line will be the shortest distance.

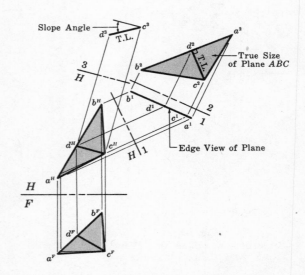

Fig. 5-5. Shortest Distance from a Point to a Line
(Plane Method)

Example: Using the same set-up as in the line method, Fig. 5-5 shows line *AB* and point *C* in both the plan and front elevation views. Connect points *A* and *B* with *C* to form plane *ABC*. Show the plane first as an edge, and then in its true size. Draw a perpendicular from point *C* to line *AB*. Label this intersection point, *D*. In view 2, *CD* is the true length of the shortest distance from point *C* to line *AB*.

Note: As in Fig. 5-4, if the slope of the shortest distance is required, then, after projecting line *CD* back to the plan view, an auxiliary elevation view should be drawn which will give both the slope of the shortest distance as well as a check for true length. The bearing of the shortest distance will be shown in the plan view.

5.4 To DRAW a LINE PERPENDICULAR to a PLANE

A. Edge-View Method

Analysis: A view showing the plane as an edge will show the true length of a line perpendicular to the plane. Since the view shows the true length of this perpendicular distance, the related view will show the perpendicular line parallel to the folding line. The point where the perpendicular line pierces the given plane will be evident in the edge view, and its location in other views is determined by simple projection.

Example: In Fig. 5-6 below the plane *ABC* and the point *X* are given in both the plan and front elevation views. The auxiliary elevation view shows the plane *ABC* as an edge. From point *X* in this view, a perpendicular line is drawn to the edge view of the plane. This perpendicular line pierces the plane at *Y*. Line *XY*, being in its true length in view 1, must appear parallel to *H*-1 in the plan view. Point *Y* is obtained in the front view by simple projection.

B. Two-View Method

Analysis: A geometric theorem tells us that, if a line is perpendicular to a plane, it is perpendicular to every line on that plane. Therefore, the required perpendicular will appear perpendicular to any true length line shown on the plane in that particular view. The pierce point of this perpendicular line can be determined by the methods shown in Article 4.1.

Example: In Fig. 5-7 below, plane *ABC* and the point *X* are given in both the plan and front elevation views. The level line *AD* is drawn in the front view and projected to the plan view where it establishes the direction of all true length lines on the plane in that view. The line from *X* must be perpendicular to this true length line. A frontal line *AE* is drawn in the plan view and is projected to the front view where it establishes the direction of all true length lines on the plane in that view. Again, the line from *X* must be perpendicular to the true length line. Since the direction of the perpendicular line to the plane is established in two views, the pierce point Y is determined by the method of Article 4.1-B.

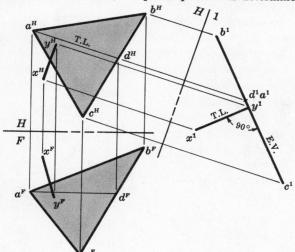

Fig. 5-6. Line Perpendicular to a Plane
(Edge-View Method)

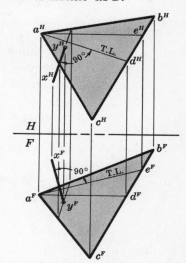

Fig. 5-7. Line Perpendicular to a Plane
(Two-View Method)

5.5 SHORTEST GRADE LINE from a POINT to a PLANE

Analysis: The plane must appear as an edge in an elevation view in order to show the slope angle of the required grade line.

Example: The plan and front elevation views of plane ABC and point X are given in Fig. 5-8 below. Draw an auxiliary elevation view showing point X and the edge view of plane ABC. To locate the shortest level distance from X to the plane, draw a line parallel to folding line H-1 from X to the edge view of the plane. To locate the shortest 30% grade line from X to the plane, draw a line from X at 30% with the H-1 line. Since the lines XD and XE are in their true length in view 1, they will appear parallel to H-1 in the plan view. Locate lines XD and XE in the front view by simple projection.

It should be noted that all shortest grade lines will have the same bearing regardless of the slope expressed.

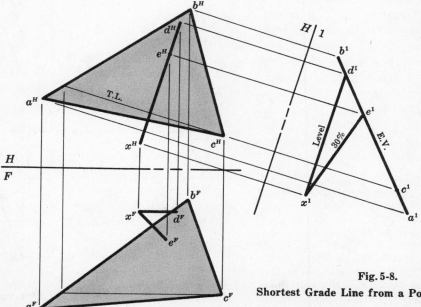

Fig. 5-8.
Shortest Grade Line from a Point to a Plane

5.6   To DRAW a PLANE through a GIVEN POINT and PERPENDICULAR to a GIVEN LINE

Analysis: This problem is the reverse of the two-view method in Article 5.4. In Fig. 5-7, the plane and the point were given, and the perpendicular line through the given point was found to be perpendicular to any true length line on the plane. In this case, however, the line is given and a point known to be in the indefinite plane is also given. Through this given point, a level line is drawn which will be in its true length and perpendicular to the given line in the plan view. Likewise, a frontal line is drawn through the given point in the plan view, and its true length will appear perpendicular to the given line in the front view. Since both lines are passed through point X, they intersect at that point, establishing a plane.

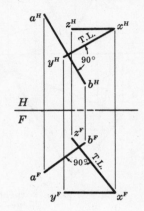

Note: An additional view showing the line in its true length will also show the plane as an edge perpendicular to the line.

Example: In Fig. 5-9, the line AB and the point X are given in both the plan and front elevation views. A level line of indefinite length YX is drawn through point X. In the plan view, the line will appear in its true length and perpendicular to the given line AB. A frontal line of indefinite length ZX is drawn through point X. In the front view the line will appear in its true length and perpendicular to the given line AB. The plane XYZ, containing the given point X, is perpendicular to the line AB.

Fig. 5-9. Plane through a Given Point and Perpendicular to a Given Line

5.7   To DRAW a PLANE through a GIVEN LINE and PERPENDICULAR to a GIVEN PLANE

Analysis: Two intersecting lines determine a plane. Since one line in the required plane is already given, the solution requires another line on the plane to be constructed. This second line which must intersect the given line also must be perpendicular to the given plane. The required plane, in order to be perpendicular to the given plane, must contain a line perpendicular to the given plane.

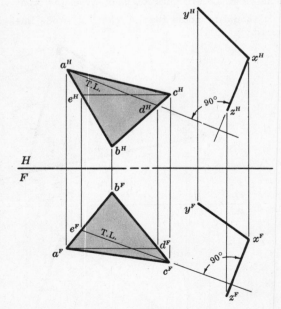

Example: In Fig. 5-10, the plane ABC and the line XY are given in both the plan and front elevation views. A level line AD is drawn and it appears in its true length in the plan view. A line XZ is drawn perpendicular to the true length line in the given plane. A frontal line CE is drawn and it appears in its true length in the front view. Again, the line XZ is drawn perpendicular to the true length line in the given plane. Since both lines pass through point X, they determine the required plane.

Fig. 5-10 Plane Containing a Given Line and Perpendicular to a Given Plane

5.8 To DRAW a PLANE through a GIVEN POINT PERPENDICULAR to EACH of TWO GIVEN PLANES

Analysis: A geometric theorem tells us that all planes containing a line perpendicular to another plane are themselves perpendicular to the other plane. If two lines, one perpendicular to one of the given planes and the other perpendicular to the other given plane, intersect at a given point, they determine a plane perpendicular to each of the two given planes.

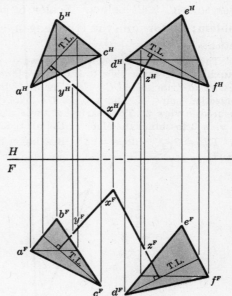

Example: In Fig. 5-11, the two planes ABC and DEF along with point X are given in both the plan and front elevation views. Level and frontal lines are shown in both given planes and they appear in their true length. In both views, line XY is passed through point X perpendicular to plane ABC. In both views, line XZ is passed through point X perpendicular to plane DEF. XYZ is the required plane which is perpendicular to both planes ABC and DEF.

Fig. 5-11. Plane through a Given Point and Perpendicular to each of Two Given Planes

Solved Problems

1. **Given:** Line AB is perpendicular to line CD. $A(4, 1, 6\frac{1}{2})$ $B(5\frac{1}{2}, 3\frac{1}{4}, 5)$ $C(3\frac{1}{2}, 2, X)$ $D(5\frac{1}{2}, 1\frac{1}{2}, 6\frac{1}{2})$. See Art. 1.7 for the coordinate system of problem layout. Refer to Fig. 5-12.

    Problem: Determine the complete coordinate location for point C. Draw the plan and any other necessary views.

    Solution:

    Using the given data, draw the front view and partial plan view. Draw an inclined view showing point D and the true length of line AB. From point D in this view, draw a line perpendicular to AB until it intersects the projection from point C in the front view. Point C is thus located in the inclined view and can now be transferred to the plan view. The missing ordinate for point C is found in the plan view. *Ans.* $C(3\frac{1}{2}, 2, 5\frac{1}{4})$

Fig. 5-12

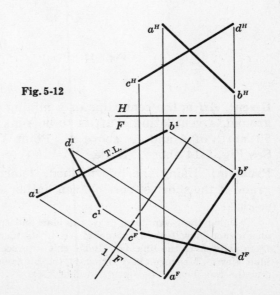

2. Given: Point $C(4, 3, 4\frac{1}{2})$ is the vertex of a triangle whose base is $A(1, 2\frac{1}{2}, 5)$ $B(3, 1, 6)$. Refer to Fig. 5-13 below.   Scale: $6'' = 1'-0''$.

Problem: Determine the true length, slope, and bearing of the altitude of the triangle.

Solution:

The plan and front elevation views of plane ABC can be determined by the given data. Draw an auxiliary elevation view to show the plane as an edge. Draw inclined view 2 which will show the true size of triangle ABC. From point C in view 2, draw a perpendicular to line AB. Label this intersection point, X, and project CX back to the other views. The true length of the altitude is measured in view 2. The bearing of the altitude is measured in the plan view. Draw auxiliary elevation view 3 to obtain the slope of the altitude. The true length will also appear in this view.

Ans. T.L. $= 2\frac{9}{16}''$, Slope $= 36°$, Bearing $= N\,56°\,W$

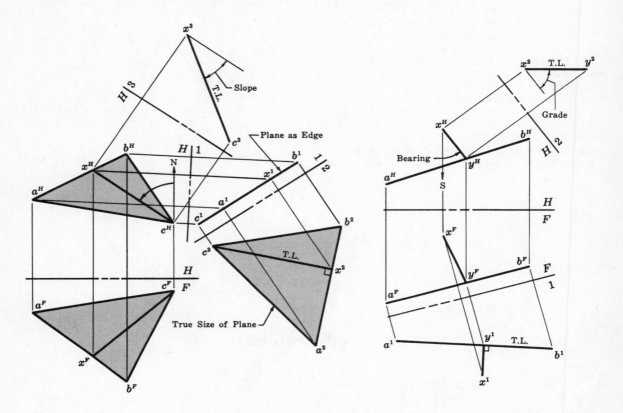

Fig. 5-13   Fig. 5-14

3. Given: AB is the centerline of a mining tunnel.   From point X on the surface of the ground, a ventilating shaft is to be sunk to the tunnel.   Point B is located 150′ east, 50′ north of A and 40′ above A.   Point X is 60′ east, 60′ north of A and 75′ above A. See Fig. 5-14 above.   Scale: $1'' = 100'$.

Problem: Using the line method, determine the true length, bearing, and per cent grade of the shortest possible ventilating shaft.   Show the ventilating shaft in all views.

Solution:

Using the given data, draw the plan and front elevation views of tunnel AB and point X. Draw an inclined view 1 showing tunnel AB in its true length and point X. Construct a perpendicular line from point X to the line AB. Label this intersection point, Y. Project line XY back to the front and plan views. Measure the bearing in the plan view. Draw an auxiliary elevation view 2 in order to find the true length and grade of line XY.   *Ans.* T.L. $= 65'-6''$, Bearing $= S\,37°\,E$, Grade $= -128\%$

4. **Given:** Two struts are determined by the lines AB and CD which are to be connected by another strut perpendicular to both AB and CD. Point B is 12' east, 16' north of A and 14' below A. Point C is located 6' west, 10' north of A and at the same elevation as A. Point D is 4' east, 12' north of A and 16' below A. Refer to Fig. 5-15 below. Scale: $\frac{3}{32}'' = 1'-0''$.

Problem: Using the line method, determine the true length, slope, and bearing of the connecting strut.

Solution:

Using the given data, locate the front and plan views of struts AB and CD. Draw an auxiliary elevation view 1 showing both struts, one of which, AB, is shown true length. Draw inclined view 2 in order to see AB as a point. In this same view, draw CD and a line perpendicular to CD from the point view of AB. This perpendicular distance is the true length of the connecting strut. Label each end of the strut X and Y. Project the strut XY back to the other views. The bearing of XY is measured in the plan view. Draw auxiliary elevation view 3 to obtain the slope of the connecting strut.

Ans. T.L. = 7'–0'', Slope = 30°, Bearing = N 77°30' W

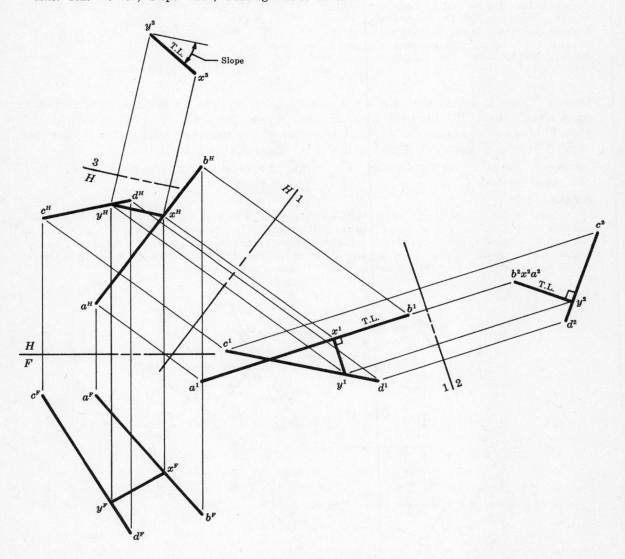

Fig. 5-15

5. Given: The base of a right pyramid, $ABCD$, is located as follows: $A(2, 2, 5)$ $B(3\frac{1}{2}, 1\frac{1}{2}, 6)$ $C(4\frac{3}{4}, 2, 6)$ $D(3\frac{1}{4}, 2\frac{1}{2}, 5)$. The altitude of the pyramid is $2\frac{1}{2}''$. See Fig. 5-16.   Scale: $\frac{1}{2}'' = 1''$.

Problem: Draw the plan and front elevation views of the pyramid showing proper visibility.

Solution:

Using the given data, draw the base of the pyramid in both the plan and front elevation views. Locate the center of the base by drawing diagonals. Label the center of the base, point X. Draw inclined view 1 to show the base plane as an edge. From point X in this view draw a perpendicular line $2\frac{1}{2}''$ long. Label the vertex of the pyramid, point Y. Project the vertex back to the other views. Connect points A, B, C, and D to point Y to complete the views. Careful visualization will determine proper visibility.

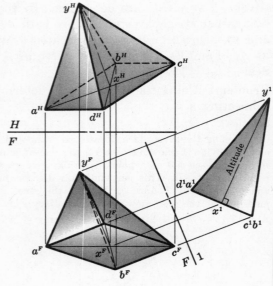

Fig. 5-16

6. Given: There is a problem of possible interference between an existing 12″ diameter pipeline AB and a 10′ diameter spherical tank. See Fig. 5-17. Point B is located 12′ east, 5′ north of A and 3′ below A. The center of the spherical tank, point X, is to be located 4′ east, 8′ north of A and 6′ below A.   Scale: $\frac{1}{8}'' = 1'-0''$.

Problem: What would be the clearance, if any, between the pipe and the tank? Use the plane method.

Solution:

Using the given data, locate the line AB and the point X in both the plan and front elevation views. Connect point X to each end of line AB to form plane ABX. In auxiliary elevation view 1, show the plane as an edge. Draw inclined view 2 which will show the true size of plane ABX. Using X as the center, draw a partial sphere. Show the pipe diameter on line AB. Measure the clearance as indicated by drawing a perpendicular from X to line AB.   *Ans.* Clearance $= 1'-9''$

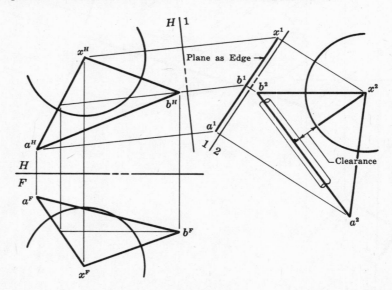

Fig. 5-17

7. **Given:** Plane ABC and point X are located as follows: B is 8′ due east of A and at the same elevation as A. Point C is 3′ east, 6′ north of A and 4′ below A. Point X is located 6′ east, 1′ north of A and 5′ below A. Refer to Fig. 5-18 below. Scale: $\frac{1}{8}'' = 1'-0''$.

   Problem: Find the shortest horizontal distance from the point X to the plane ABC. Determine the length of the shortest line from X to the plane and having a slope of 30°. What is the bearing of each line? Show both lines from X in all views.

   Solution:

       Draw the plan and front elevation views of plane ABC and point X. Since line AB is a frontal-level line, the plane will appear as an edge in the profile view. Draw a level line from point X to the plane ABC extended. Label the intersection point Y and project XY back to the other views. In the profile view also, draw a line from X to intersect the plane and sloping 30°. Label the intersection point Z and project XZ back to the other views. Line XY is the shortest horizontal distance from X to plane ABC and line XZ is the shortest line from X to ABC, having a slope of 30°. The true lengths are measured in the profile view and the bearings are measured in the plan view.

   Ans. Level T.L. = $6'-6\frac{1}{2}''$, 30° Slope Line T.L. = $4'-0''$, Bearing XY and XZ = Due North-South

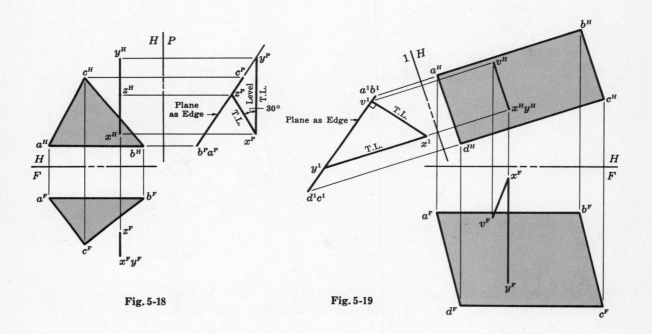

Fig. 5-18   **Fig. 5-19**

8. **Given:** A television antenna extends above the roof of a house as shown in Fig. 5-19 above. The roof plane $ABCD$ and the top of the antenna, point X, are located as follows: $A(3, 3\frac{1}{2}, 6\frac{1}{2})$ $B(6, 3\frac{1}{2}, 7\frac{1}{2})$ $C(6\frac{1}{2}, 1\frac{1}{2}, 6)$ $D(3\frac{1}{2}, 1\frac{1}{2}, 5)$ $X(4\frac{1}{2}, 4\frac{1}{4}, 5\frac{3}{4})$. Scale: $\frac{1}{8}'' = 1'-0''$.

   Problem: If a supporting brace extending to point X is perpendicular to the roof, how long must the brace be? How long is the vertical antenna? Show the brace and antenna in all views.

   Solution:

       Using the given data, lay out the plan and front elevation views of the roof plane and the top of the antenna. Draw auxiliary elevation view 1 to show point X and the edge view of the roof plane. From point X in view 1, draw a perpendicular to the edge view of the plane $ABCD$. This distance will be the true length of the brace. Label the intersection of the brace and roof as point V and project it back to the plan and front elevation views. From point X in view 1, draw a vertical line perpendicular to the H-1 folding line. The intersection of this line with the roof will be the true length of the antenna. Label the intersection point, Y, and project it back to the plan and front elevation views. *Ans.* Brace T.L. = $5'-7''$, Antenna T.L. = $9'-0''$

9. **Given:** Plane ABC and point X are shown in Fig. 5-20 below. The plane and point X are located as follows: $A(4, 3, 4\frac{1}{2})$ $B(5, 1, 5\frac{1}{2})$ $C(6, 2\frac{1}{2}, 4)$ $X(5\frac{1}{2}, 3, 5)$. See Art. 1.7 for the coordinate system of problem layout. Scale: $\frac{1}{8}'' = 1'-0''$.

Problem: Determine the true length and bearing of the shortest line from X to the plane and having a $-45°$ slope. Use the edge-view method.

Solution:

Draw the plan and front elevation views of plane ABC and point X as per given data. Draw auxiliary elevation view 1 to show point X and the edge view of plane ABC. From point X, draw a line at $-45°$ slope until it intersects the edge view of the plane. Label the intersection point, Y, and project line XY back to the plan and front views, using proper visibility. The true length of the required line is measured in view 1 and the bearing of XY is measured in the plan view.

Ans. T.L. $= 3'-7''$, Bearing $= $ S $24°$ W

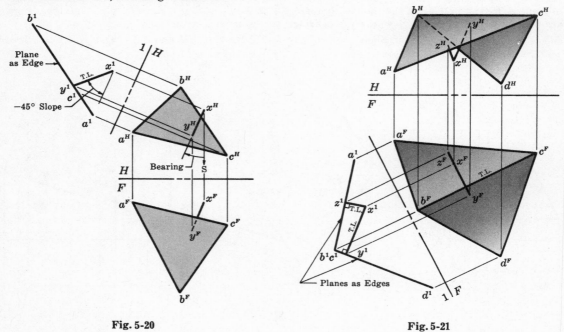

Fig. 5-20   **Fig. 5-21**

10. **Given:** Point X and intersecting planes ABC and BCD, having BC as the common line of intersection. See Fig. 5-21 above. Point B is 2' east, 5' north of A and 6' below A. Point C is 12' east of A, 5' north of A and 1' below A. Point D is 9' east, 1' south of A and 10' below A. Point X is 5' east, 1' north of A and 2' below A. Scale: $\frac{1}{8}'' = 1'-0''$.

Problem: How far is point X from each plane? Show these distances in all views with proper visibility.

Solution:

Using the given data, draw the plan and front elevation views of planes ABC, BCD, and point X. Since the common line of intersection, BC, is a frontal line, it will appear true length in the front elevation view. Locate folding line F-1 perpendicular to the true length line of intersection and project to secure the edge view of each plane and point X. Draw a perpendicular to each plane from point X. The perpendicular distance is the true distance from the point to each plane. Project the lines XY and XZ back to the other views. Use careful visualization to determine proper visibility.

Ans. $XY = 4'-6''$, $XZ = 1'-9''$

11. **Given:** Plane $ABCD$ and points X and Y. The plane and points are located as follows: $A(1, 2, 5\frac{1}{2})$ $B(2, 2\frac{1}{2}, 4)$ $C(4, 1, 5)$ $D(3, \frac{1}{2}, 6\frac{1}{2})$ $X(2, 3\frac{1}{4}, 6\frac{1}{2})$ $Y(1\frac{1}{4}, 1, 4\frac{1}{2})$. See Fig. 5-22 below. Scale: $\frac{1}{8}'' = 1'-0''$.

Problem: Which of the points is closest to the plane and by how much? Show the perpendicular distances in all views. Determine proper visibility.

Solution:

Using the given data, locate the plan and front elevation views of the plane *ABCD* and the points *X* and *Y*. Draw auxiliary elevation view 1 to show points *X* and *Y* in relation to the edge view of plane *ABCD*. Draw a perpendicular line from both *X* and *Y* to the edge view of the plane. The perpendicular distance is the true distance from each point to the plane. Measure both true length distances and subtract to determine which point is closer and by how much. Label the intersection of the line from *X* to the plane with the letter *V*. Label the intersection of the line from *Y* to the plane with the letter *Z*. Project lines *XV* and *YZ* back to the other views.

Ans. *Y* is closer by 2′–7″

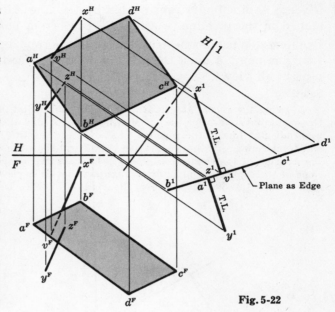

Fig. 5-22

12. **Given:** *ABC* and *DEF* are parallel planes and point *X* is located on the plane *ABC*. Point *B* is 7′ east, 5′ north of *A* and 4′ below *A*. Point *C* is located 8′ east, 2′ south of *A* and at the same elevation as *A*. Point *D* is 1′–6″ east, 1′ south of *A* and 5′ above *A*. Point *E* is located 8′–6″ east, 4′ north of *A* and 1′ above *A*. Point *F* is located 9′–6″ east, 3′ south of *A* and 5′ above *A*. The point *X* on plane *ABC* is 7′ due east of *A*. See Fig. 5-23 below. Scale: $\frac{1}{8}'' = 1'-0''$.

Problem: Determine the shortest distance from point *X* to the plane *DEF*. Show this distance in all views. Show proper visibility.

Solution:

Draw the plan and front elevation views of the two planes. Locate point *X* in the plan view. Draw auxiliary elevation view 1 to show both planes as parallel edge views. Project point *X* from the plan to the auxiliary elevation view 1, locating it on the edge view of plane *ABC*. Draw a perpendicular from *X* to the edge view of plane *DEF*. Label this point, *Y*, and project the line *XY* back to the plan and front views. In the plan view, the line *XY* will be parallel to folding line *H*-1, since the true length of the required line is shown in view 1. Use careful visualization to determine proper visibility.

Ans. Distance = 4′–0″

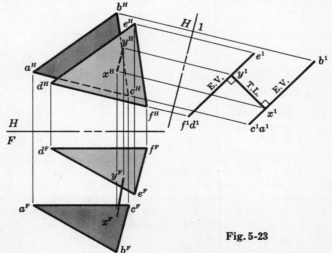

Fig. 5-23

13. Given: A pipeline is represented by the line AB in Fig. 5-24. A 90° tee is to be installed in the pipeline to connect a pipe from point X. Scale: $\frac{1}{16}'' = 1'-0''$.

Problem: What is the true length, slope, and bearing of the connecting pipe? How far from point A must the pipeline be cut? Use the line method. Show the connecting pipe in all views.

Solution:

Using the given data, locate the pipeline and point X in both the plan and front elevation views. Draw auxiliary elevation view 1 showing point X and the true length of AB. From point X, draw a perpendicular line to the pipeline. Label this connecting point, Y, and project XY back to the plan and front views. Auxiliary elevation view 1 will show the true distance of the connection from point A. The bearing of the connecting pipe will be measured in the plan view. To obtain the true length and slope of the connecting pipe, draw auxiliary elevation view 2.

Ans. T.L. = 13'-5'', Slope = 53°, Bearing = N 26° E, Distance from A = 7'-6''

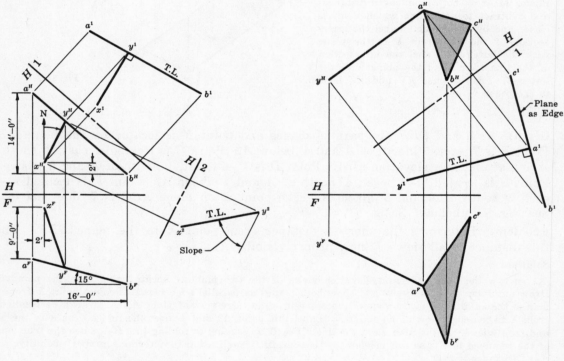

Fig. 5-24   Fig. 5-25

14. Given: Plane ABC and point Y located as follows: $A(3, 2, 8)$ $B(3\frac{1}{2}, X, 6\frac{1}{2})$ $C(4, 3\frac{1}{2}, X)$ $Y(1, 3, 6\frac{1}{2})$. Refer to Fig. 5-25.

Problem: Draw the plane ABC perpendicular to line YA. What is the approximate coordinate location of point B?

Solution:

Using the given data, draw partial plan and front elevation views of the plane and point Y. Connect point Y to point A in both views. Draw auxiliary elevation view 1 to show line YA in its true length. At point A in view 1 draw a line perpendicular to the true length of YA. Project point B from the plan view to view 1 and locate point C in this view also. Both points B and C will be located on the line drawn perpendicular to the true length of YA. Points A, B, and C in view 1 establish the edge view of the plane which is perpendicular to the line YA. Complete the plan and front elevation views.   *Ans.* $B(3\frac{1}{2}, \frac{3}{4}, 6\frac{1}{2})$

15. Given: Planes ABC, DEF and point X located as follows: Point B is 2′ east, 4′ north of A and 2′ above A. Point C is 5′ east, 2′ north of A and 3′ below A. Point D is 9′ east, 2′ north of A and 4′ below A. Point E is 12′ east, 5′ north of A and 2′ above A. Point F is 14′ east, 1′ south of A and 1′ below A. Point X is located 7′ east, 3′ south of A and 3′ above A. See Fig. 5-26 below. Scale: $\frac{3}{16}'' = 1'-0''$.

Problem: Locate a plane through point X which is perpendicular to both planes ABC and DEF. Determine the slope of this new plane.

Solution:

Using the given data, draw the plan and front elevation views of planes ABC, DEF and point X. Draw a true length line in each of the given planes in both views. In both views, line XY is passed through point X perpendicular to plane ABC. In both views, line XZ is passed through point X perpendicular to plane DEF. XYZ is the required plane which is perpendicular to both planes ABC and DEF. Auxiliary elevation view 1 will show the plane XYZ as an edge and the slope can be measured in this view.   *Ans.* Slope $= 46°$

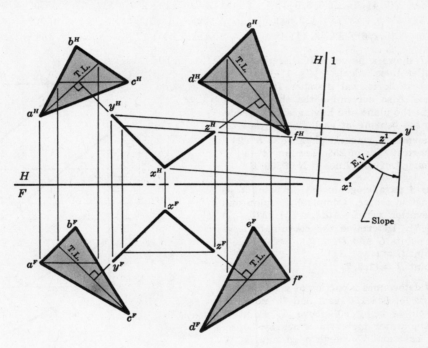

Fig. 5-26

Supplementary Problems

In each of the following problems 16 to 25 determine the true length, slope, and bearing of the shortest distance from point X to line AB. Show the shortest distance as a line in all views. Measure the true length to the nearest sixteenth of an inch. Scale: $12'' = 1'-0''$.

See Art. 1.7 for the coordinate system of problem layout.

				T.L.	Slope	Bearing
16.	$A(2,4,5)$	$B(4,2,6\frac{1}{2})$	$X(3,3,5\frac{1}{2})$	Ans. $\frac{1}{4}''$	$16°$	N 22° W
17.	$A(2,3\frac{1}{2},4\frac{1}{2})$	$B(2,2,6)$	$X(2\frac{1}{2},1\frac{1}{2},5)$	Ans. $1\frac{3}{16}''$	$39°30'$	N 34° W
18.	$A(2,4,6)$	$B(4,4,5)$	$X(3,3\frac{1}{4},5\frac{1}{4})$	Ans. $\frac{13}{16}''$	$73°$	N 26° E
19.	$A(2,4\frac{1}{2},6)$	$B(4,4\frac{1}{2},6)$	$X(2\frac{1}{2},3,5)$	Ans. $1\frac{13}{16}''$	$56°$	Due N–S
20.	$A(2,3,6)$	$B(2,2,4\frac{1}{2})$	$X(2,3\frac{1}{2},5)$	Ans. $\frac{15}{16}''$	$56°30'$	Due N–S
21.	$A(1,2\frac{1}{2},5)$	$B(4,4,5\frac{1}{4})$	$X(2\frac{1}{4},2,6)$	Ans. $1\frac{3}{8}''$	$42°$	S 23° W
22.	$A(2,1,4\frac{1}{2})$	$B(4\frac{1}{2},3\frac{1}{2},6)$	$X(3,2\frac{1}{2},6\frac{1}{2})$	Ans. $1\frac{3}{16}''$	$3°30'$	S 28°30' E
23.	$A(5,1\frac{1}{2},5)$	$B(5,4\frac{1}{4},6)$	$X(3,3,5\frac{1}{4})$	Ans. $2''$	$3°30'$	N 82° E
24.	$A(2\frac{1}{4},1,5)$	$B(3\frac{1}{2},4,6)$	$X(2\frac{1}{2},3,6\frac{1}{2})$	Ans. $1\frac{1}{16}''$	$1°30'$	S 35°30' E
25.	$A(2,4\frac{1}{4},6)$	$B(5,2,5)$	$X(5,3,4\frac{1}{2})$	Ans. $1''$	$45°$	N 29° W

26. Fig. 5-27 shows a plane, $ABCD$, represented by parallel lines. Scale: $\frac{1}{4}'' = 1'-0''$. Find the shortest horizontal distance from X to the plane. Also determine the shortest line from X to the plane and having a 25% grade. What is the bearing of each line?

Ans. Shortest Horizontal distance $= 5'-2\frac{1}{2}''$
Shortest Line of 25% Grade $= 4'-3''$
Bearing of both lines $=$ N 65°30' E

27. $ABCD$ is a parallelogram having point X as the geometric center. Complete the plan and front elevation views. $A(2\frac{1}{2},4,5\frac{1}{2})$ $B(5,3,7)$ $X(3\frac{1}{2},2\frac{1}{2},7)$. Determine the coordinate locations of points C and D.
Ans. Point $C = (4\frac{1}{2},1,8\frac{1}{2})$
Point $D = (2,2,7)$

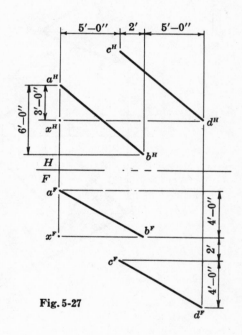

Fig. 5-27

28. Line AB determines a portion of a water line. Point B is located 150' east and 75' north of A, as well as being 50' above A. Point X locates the water meter in a nearby house. X is 75' east and 75' north of A and is 75' above A. Scale: $1'' = 50'$. Determine the shortest distance from the meter to the main line if a 45° connection is made at the main line. Ans. T.L. $= 76'$

29. In Fig. 5-28 we have shown the plan of a corner lot in a residential area. Scale: $1'' = 100'$. AB and BC are the centerlines of an existing water main. If the water meter in the house is located at D, what would be the shortest distance to the existing water line? Show this shortest distance in all views.
Ans. 155'

30. What would be the true length of a branch pipe from X to the line AB if the connection is to be made with a 90° tee? Scale: $1'' = 40'$. $A(1,1,4)$ $B(3\frac{1}{2},2,4)$ $X(1,2\frac{1}{2},4)$. Determine the slope and bearing of this new connecting pipe.
Ans. T.L. $= 55'-6''$, Slope $= -68°$
Bearing $=$ Due East–West

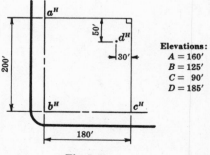

Fig. 5-28

Elevations:
A = 160'
B = 125'
C = 90'
D = 185'

31. AB is a pipe segment to which another pipe from C must be connected with a 45° elbow. $A(1,1\frac{1}{2},8)$ $B(3\frac{1}{2},3\frac{1}{2},8)$ $C(2,1,9)$. Scale: $\frac{1}{4}'' = 1'-0''$. Determine the true length, slope, and bearing of the connecting pipe. Ans. T.L. $= 8'-2''$, Slope $= 56°30'$, Bearing $=$ S 27° E

32. Fig. 5-29 shows the roof of a cottage and the top of a nearby flagpole. Scale: $1'' = 10'-0''$. The shortest possible guy wire is to be anchored to the roof from the top of the pole. How long must this guy wire be?   *Ans.* T.L. $= 17'-10''$

33. *AB* represents a coal mining tunnel. An explosion has sealed off the tunnel at the main access shaft from the surface of the earth. It is decided to start a new shaft from point *X* on the earth's surface in order to reach the entombed miners. Point *B* is located 150′ east, 50′ north of *A* and 40′ above *A*. Point *X* is 60′ east, 60′ north of *A* and 75′ above *A*. Scale: $1'' = 50'$. Determine the true length, bearing, and grade of the emergency shaft. If the disaster crew averages 13′ per hour, how long will it take them to reach the tunnel which has been sealed off from the main shaft?
Ans. T.L. $= 65'-6''$
   Bearing $=$ S 37° E
   Grade $= 128\%$
   Time $=$ About 5 hrs.

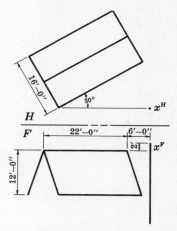

Fig. 5-29

In each of the following problems 34 to 38 determine the true length of the shortest distance from point *X* to the plane *ABC*. What is the bearing of the shortest distance?   Scale: $1'' = 1'-0''$.

See Art. 1.7 for the coordinate system of problem layout.

				T.L.	Bearing
34. $A(2,3,5)$ $B(3,1,6\frac{1}{2})$ $C(3\frac{1}{2},2\frac{1}{2},4)$ $X(2\frac{1}{2},\frac{1}{2},4\frac{1}{4})$			*Ans.*	$1'-9''$	N 47° E
35. $A(1,3,6)$ $B(2\frac{1}{4},2,6)$ $C(4,4,5\frac{1}{4})$ $X(3,4\frac{1}{4},7)$			*Ans.*	$1'-6\frac{1}{2}''$	S 10°30′ W
36. $A(1,1\frac{1}{2},5\frac{1}{2})$ $B(3,\frac{1}{2},4)$ $C(4\frac{1}{2},3,6)$ $X(2,\frac{1}{2},6\frac{1}{2})$			*Ans.*	$1'-6\frac{3}{4}''$	S 14°30′ W
37. $A(1\frac{1}{2},3,4)$ $B(2,3,5)$ $C(4\frac{1}{2},3\frac{1}{2},6)$ $X(3\frac{1}{2},3,7\frac{1}{4})$			*Ans.*	$0'-1\frac{1}{4}''$	N 63°30′ W
38. $A(2,4,7)$ $B(2,3,5)$ $C(4,3,4)$ $X(3\frac{1}{4},1,5)$			*Ans.*	$1'-11\frac{3}{4}''$	S 27° W

39. Fig. 5-30 shows a transition piece which is to be connected by a cable passing through point *X*. Scale: $\frac{1}{2}'' = 1'-0''$. Determine the shortest possible distance from the point *X* to the nearest face of the transition.   *Ans.* Distance $= 2'-1\frac{1}{2}''$

40. Plane *ABC* and point *X*. Point *B* is 10′ east, 8′ north of *A* and 12′ below *A*. Point *C* is 16′ east, 2′ south of *A* and 5′ below *A*. Point *X* is 6′ east, 2′ south of *A* and 14′ below *A*. Scale: $\frac{1}{8}'' = 1'-0''$. Determine the shortest horizontal distance from point *X* to the plane of *ABC*. What would be the shortest line from *X* to the plane having a grade of 35%? Determine the bearing of each line from *X*.
Ans. Shortest Horizontal Distance $= 12'-7''$
   Shortest line having 35% Grade $= 10'-0''$
   Bearing of both lines $=$ N 23°30′ E

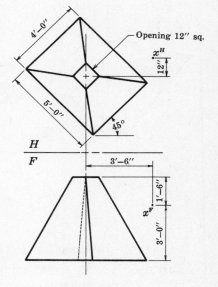

41. Line *AB* determines a portion of a water line. Point *B* is located 150′ east and 75′ north of *A*, as well as being 50′ above *A*. Point *X* locates the water meter in a nearby house. *X* is 75′ east and 75′ north of *A* and is 75′ above *A*. Scale: $1'' = 50'$. Determine the true length, slope, and bearing of the shortest possible connecting pipeline from the water meter to the main water line. Solve by using the line method. Show the connecting pipeline in all views.
Ans. T.L. $= 54'$
   Slope $= -50°$
   Bearing $=$ S 46° 30′ E

Fig. 5-30

42. Two lines, *AB* and *CD*, intersect to form a plane. It is required to find the distance from a given point *X* to the plane of *ABCD*. Point *B* is located 9′ east, 5′ south of *A* and its elevation is not given. Point *C* is 1′ west, 8′ south of *A* and 5′ below *A*. Point *D* is 6′ east, 1′ north of *A* and 1′ above *A*. The point *X* is located 7′ east, 3′ south of *A* and 2′ above *A*. Scale: $\frac{1}{4}'' = 1'-0''$.
Ans. Distance $= 2'-10''$

Chapter 6

Dihedral Angle and
Angle Between Line and Plane

An engineer is often required to determine the angle between two intersecting planes. He is also expected to know how to determine the true size of the angle between a line and a plane. The knowledge of the underlying principles involved in these two items may be considered essential to most students of engineering and science.

6.1 DIHEDRAL ANGLE

Definition: A dihedral angle is the angle formed by two intersecting planes. The dihedral angle is measured in a plane perpendicular to the line of intersection of the two planes. (See Fig. 6-1 below.)

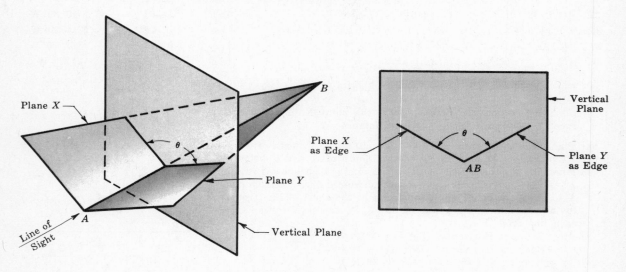

Fig. 6-1. Dihedral Angle

A. Line of Intersection Given

Analysis: A view which shows the line of intersection of the two planes as a point will also show the two planes as edges. The angle between the two edge views of the planes is the true size of the dihedral angle.

Example: (Fig. 6-2 below.) It is desired to find the dihedral angle between the two given planes, *ABD* and *ACD*, having the line of intersection of the two planes given. An auxiliary elevation view is drawn showing the line of intersection *AD* in its true length. The inclined view 2 is then drawn to show the line of intersection as a point. This view also shows the two given planes as edges. The dihedral angle is measured between the two edge views.

Since inclined view 2 shows the edge views of both planes, an additional view projected from each edge view will show the true size of each plane. Inclined view 3 shows the true size of plane *ACD*.

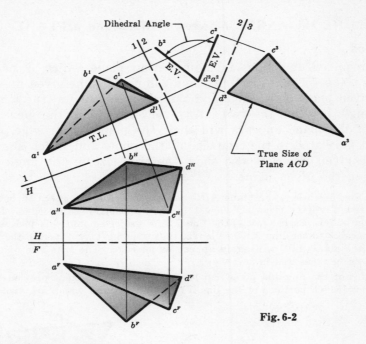

Fig. 6-2

B. Line of Intersection Not Given

Analysis: If both planes can be seen as edges in the same view the dihedral angle can be measured between the two edge views.

Example: In Fig. 6-3 below, the two planes *ABC* and *DEF* are given in both the plan and front elevation views. An auxiliary elevation view is drawn showing plane *DEF* as an edge. Plane *ABC* is also projected into each view. An inclined view 2 is then drawn to show the true size of plane *DEF*. Any view that is projected from view 2 will show the plane *DEF* as an edge. However, in order for plane *ABC* to appear as an edge in the inclined view 3, a line on plane *ABC* must be shown as a point in that view. Therefore, in view 1 a line parallel to folding line 1-2 is drawn from *B* to *Y* so that line *BY* will appear in its true length in view 2. This true length line in view 2 establishes the line of sight for inclined view 3 because it will show line *BY* as a point on the edge view of plane *ABC*. Either angle formed by the two edge views can be considered the dihedral angle.

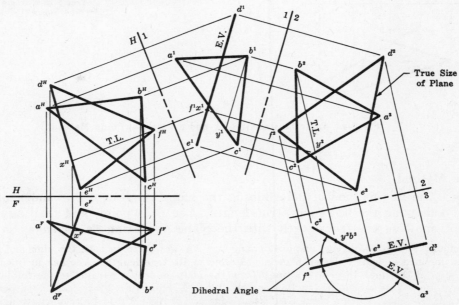

Fig. 6-3

6.2 To DETERMINE the ANGLE between a LINE and a PLANE

A. Plane Method

Analysis: The required angle lies in a projecting plane which is perpendicular to the given plane. The projecting plane will contain the given line. A view showing the given plane as an edge and the given line in its true length will also show the angle between the two. Any view related to the view in which the true size of the plane appears will show the plane as an edge. Therefore a view which is related to the true size of the plane and which shows the true length of the given line will also show the true size of the angle made by the line piercing the given plane.

Example: In Fig. 6-4 below, the plane *ABC* and the line *XY* are given in both the plan and front elevation views. Draw auxiliary elevation view 1 to show the plane as an edge. Inclined view 2 shows the true size of plane *ABC*. As in view 1, the line *XY* is projected into this inclined view also. Folding line 2-3 is located parallel to the line *XY* in view 2, thus yielding view 3 where the line will be shown in its true length and the plane *ABC* will again appear as an edge. The true size of the angle between the line and the plane can be measured in this view.

The pierce point of the line and plane can be determined by methods previously explained. Careful visualization will show which portions of the line *XY* are hidden and which are visible.

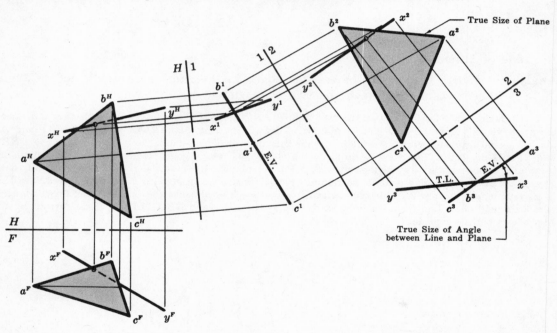

Fig. 6-4. Angle between a Line and a Plane — (Plane Method)

B. Line Method

Analysis: The line method of determining the angle between a line and a plane involves showing the line in its true length, then as a point, and then again in a view showing the true length with the plane appearing as an edge.

Example: In Fig. 6-5 below, plane *ABC* and line *XY* are given in both the plan and front elevation views. An inclined view 1 is drawn showing the line in its true length. Another inclined view 2 shows the point view of the line. In view 1 a line *BD* on the plane is drawn parallel to the folding line 1-2 thus appearing in view 2 as a true length line on the plane. Folding line 3-2 is located perpendicular to the true length of *BD*, thus showing plane *ABC* as an edge and the line *XY* in its true length again. The intersection of the line and the plane in this view determine the true size of the angle between them.

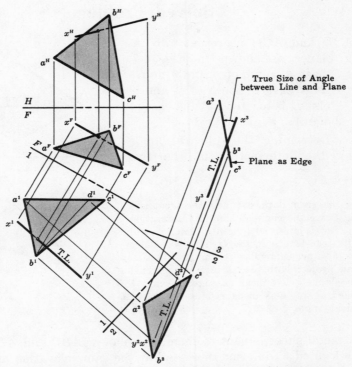

Fig. 6-5. Angle between a Line and a Plane — (Line Method)

C. Complementary-Angle Method

Analysis: If the complement of the required angle can be determined, then the angle itself will be known.   Therefore, select any point on the given line and draw a perpendicular to the given plane.   This perpendicular line and the given line determine a plane.   Show the true size of this newly-constructed plane, and the required angle will be 90° minus the true size of the complementary angle.

Example: In Fig. 6-6 the plane ABC and the line XY are given in both the plan and front elevation views. Draw a true length line in both the plan and front elevation views of the plane ABC. From point X in each view draw a line of indefinite length, XZ, perpendicular to the true length lines. Plane ABC can now be disregarded since it is immaterial where line XZ pierces the plane. Draw an auxiliary elevation view showing plane XYZ as an edge. Draw an inclined view 2 showing plane XYZ in its true size. The complementary angle YXZ is seen in its true size in this view. Construct a 90° angle at point X which includes the complementary angle. The right angle minus the complementary angle will determine the required angle between the line XY and the plane ABC.

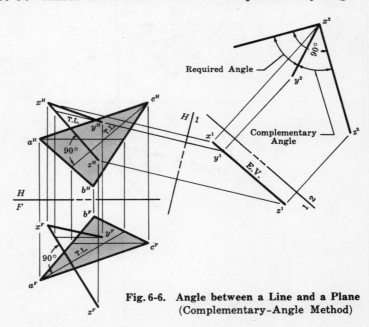

Fig. 6-6. Angle between a Line and a Plane
(Complementary-Angle Method)

Solved Problems

1. **Given:** *ABD* and *ACD* represent two intersecting planes as shown in Fig. 6-7. Point *B* is 6' east, 4' north of *A* and 6' above *A*. Point *C* is 6' east, 3' south of *A* and 6' above *A*. Point *D* is located 10' due east of *A* and 3' above *A*. Scale: $\frac{1}{8}'' = 1'-0''$.

   Problem: Determine the dihedral angle formed by the intersection of the two planes.

   Solution:

   Using the given data, complete the front and plan views. Since the common line of intersection between the two planes is a frontal line, it will appear in its true length in the front view. Place folding line *F*-1 perpendicular to the intersection line *AD* and project both planes into the inclined view. Both planes will appear as edges in view 1 and the dihedral angle is measured in this view. *Ans.* Dihedral Angle = 82°

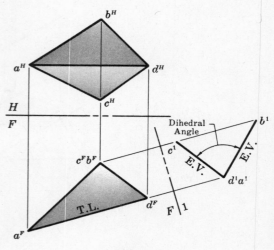

Fig. 6-7

2. **Given:** A metal sheet is bent to form two planes, *ABC* and *ABD*. See Fig. 6-8 below. Point *B* is located 6' due east of *A* and at the same elevation as *A*. Point *C* is 8' east, 2' south of *A* and 2' above *A*. Point *D* is 3' east, 3' south of *A* and 3' below *A*. Scale: $\frac{3}{16}'' = 1'-0''$.

   Problem: Determine the dihedral angle between the two planes. Show the true size of each plane.

   Solution:

   Draw the plan and front elevation views of both planes. Since the common line of intersection between the two planes is a frontal-level line, a profile view will show the line of intersection as a point and both planes as edges. The dihedral angle is measured in the profile view. Draw inclined views 1 and 2 in order to obtain the true size of each plane.   *Ans.* Dihedral Angle = 90°

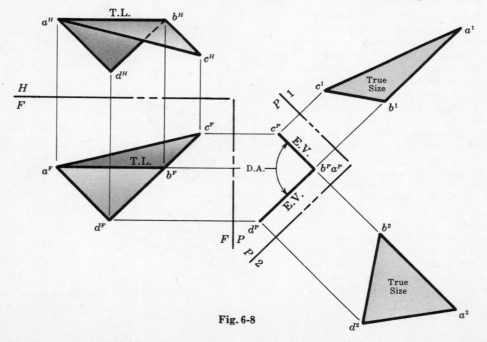

Fig. 6-8

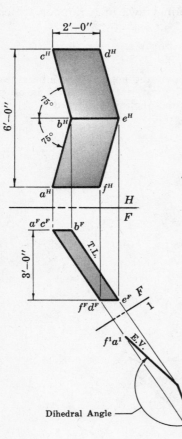

Fig. 6-9

3. **Given:** Plan and front elevation views of an airplane windshield section. See Fig. 6-9. Scale: $\frac{1}{4}'' = 1'-0''$.

Problem: Determine the dihedral angle between the two window panes.

Solution:

Since the metal frame *BE* is a frontal line, it will appear in its true length in the front view. Draw an inclined view showing the line of intersection as a point and the two planes as edges. The dihedral angle is measured in this view.
Ans. Dihedral Angle = 155°

4. **Given:** Fig. 6-10 shows partial plan and front elevation views of a hip roof. Scale: $\frac{1}{16}'' = 1'-0''$.

Problem: Determine the dihedral angle between planes *A* and *B*.

Solution:

Locate a point on plane *B*, such as *D*. The two planes involved are *ABC* and *ABD*. Draw auxiliary elevation view 1 showing both planes and having the line of intersection *AB* shown in its true length. Draw inclined view 2 showing the line of intersection as a point and the planes as edges. The angle between the two edge views is the dihedral angle.   *Ans.* Dihedral Angle = 119°30′

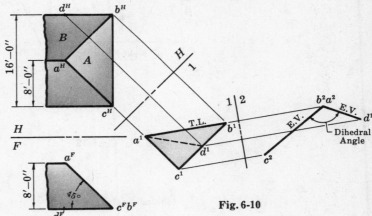

Fig. 6-10

5. **Given:** Two planes, *ABC* and *ABD*, intersect along line *AB*. Point *B* is 4′ east, 8′ south of *A* and 8′ below *A*. Point *C* is 11′ east, 2′ north of *A* and 10′ below *A*. Point *D* is located 7′ due north of *A* and 11′ below *A*. See Fig. 6-11. Scale: $\frac{1}{16}'' = 1'-0''$.

Problem: Determine the dihedral angle between the two planes.

Solution:

Using the given data, draw the plan and front elevation views of both planes. Draw auxiliary elevation view 1 showing both planes and having the line of intersection shown in its true length. Construct an inclined view showing both planes as edges and the line of intersection as a point. The dihedral angle is measured in this view.   *Ans.* Dihedral Angle = 50°

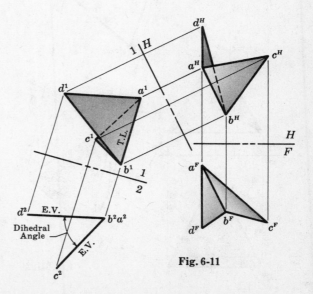

Fig. 6-11

6. Given: Fig. 6-12 below shows the plan and front elevation views of a hopper which is fed by a pipe represented by the centerline XY. Scale: $\frac{3}{16}'' = 1'-0''$.

Problem: Determine the angle the pipe makes with the hopper.

Solution:

Draw inclined view 1 to show the line XY and the true size of the hopper face intersected by the pipe. Place folding line 1-2 parallel to the line XY in view 1. Draw inclined view 2 in order to obtain the true length of XY and the plane as an edge. The required angle is measured in this view. *Ans.* Angle $= 21°30'$

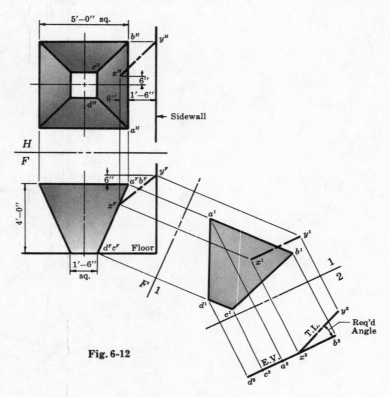

Fig. 6-12

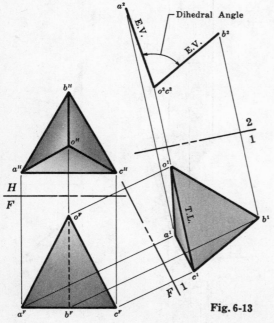

Fig. 6-13

7. Given: A triangular pyramid 2″ high has an equilateral base 2″ on a side as shown in Fig. 6-13. Scale: $\frac{1}{2}'' = 1''$.

Problem: Determine the dihedral angle between any two lateral surfaces.

Solution:

Using the given data, draw the plan and front elevation views of the pyramid. Draw an inclined view to show the common line of intersection in its true length. Locate one additional point on each of the two planes in this same view. Draw inclined view 2 in order to obtain the point view of the line of intersection between the two planes. The two planes will appear as edges in this view and the angle between the edge views will be the dihedral angle. *Ans.* Dihedral Angle $= 68°$

8. **Given:** A bulkhead is determined by the plane *ABCD* through which a control cable *XY* must pass. Point *B* is 1' west, 2' north of *A* and 3' below *A*. Point *C* is 4' west, 1' south of *A* and 6'' below *A*. Point *D* is 5' west, 1' north of *A* and 3'—6'' below *A*. Point *X* is 1' due west of *A* and 6'' above *A*. Point *Y* is 3'—6'' west, 9'' north of *A* and 4' below *A*. See Fig. 6-14 below. Scale: $\frac{1}{4}$'' = 1'—0''.

Problem: Use the plane method to determine the angle between the cable and the bulkhead. Show proper visibility.

Solution:

Draw the line *XY* and the plane *ABCD* in both plan and front elevation views. Draw auxiliary elevation view 1 to show line *XY* and the edge view of plane *ABCD*. Inclined view 2 will show the true size of the plane. Place folding line 2-3 parallel to *XY* in view 2 and project both the line and plane into inclined view 3. View 3 will show the cable *XY* in its true length and the bulkhead *ABCD* as an edge. The angle between the cable and bulkhead is measured in this view. Careful visualization will determine proper visibility.   *Ans.* Angle = 16°30'

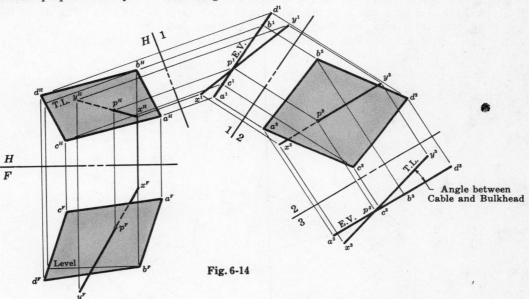

Fig. 6-14

9. **Given:** The base of a right rectangular pyramid measures 2'' long by $1\frac{1}{2}$'' wide. The vertex of the pyramid is located $2\frac{1}{2}$'' above the base but the pyramid is truncated 1'' above the base at an angle of 30°. See Fig. 6-15. Scale: $\frac{1}{2}$'' = 1''.

Problem: Determine the dihedral angle between the cut surface and the front lateral surface.

Solution:

Using the given data, construct the plan and front elevation views of the truncated pyramid. Draw inclined view 1 which shows the common line of intersection *CD* in its true length. Draw inclined view 2 which shows both planes as edges. The angle between the edge views is measured as the dihedral angle.   *Ans.* Angle = 105°

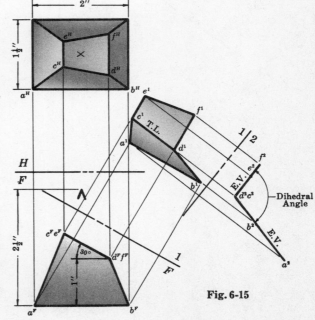

Fig. 6-15

10. Given: A mine tunnel *XY* is heading in the direction of a vein of coal which is determined by plane *ABC*. Point *B* is 40′ east, 40′ north of *A* and 60′ below *A*. Point *C* is 65′ east, 30′ south of *A* and 20′ below *A*. Point *X* is 80′ east, 40′ north of *A* and 20′ above *A*. Point *Y* is 55′ east, 20′ north of *A* and at the same elevation as *A*. See Fig. 6-16 below. Scale: 1″ = 60′.

Problem: At what angle will the mine tunnel meet the vein of coal? How much farther must the mine tunnel be extended to meet the vein of coal? If point *A* is at an elevation of 500′, at what elevation would the tunnel reach the vein of coal?

Solution:

Using the given data, draw the plan and front elevation views of plane *ABC* and the line *XY*. Draw auxiliary elevation view 1 to show the plane as an edge. Draw inclined view 2 which will show the true size of plane *ABC* and the line *XY*. Locate folding line 2-3 parallel to line *XY* in view 2 and project into another inclined view which will show the true length of line *XY* and the plane as an edge. Extend line *XY* in view 3 until it intersects the edge view of the plane. The extension of the tunnel and the angle it makes with the plane are both measured in this view. The elevation of point *P* can be measured in either elevation view 1 or the front elevation view.

Ans. Angle = 74°30′, Tunnel Extension = 38′, Elevation Point *P* = 480′

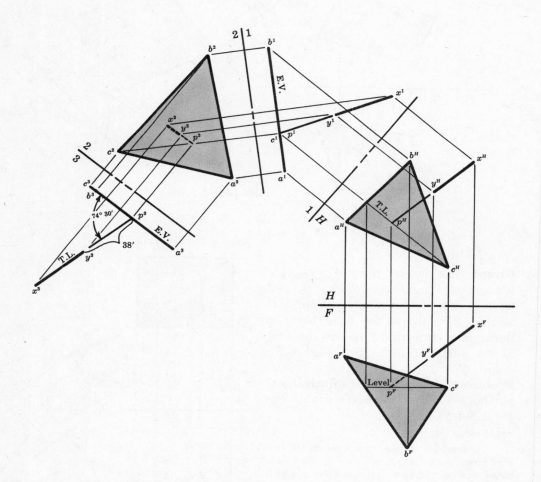

Fig. 6-16

11. Given: Two planes *ABCD* and *ABEF* are located as follows: $A(6, 2\frac{1}{2}, 8)$ $B(4\frac{1}{2}, \frac{1}{2}, 6)$ $C(7, 3, 8)$ $D(5\frac{1}{2}, 1, 6)$ $E(5, 2\frac{1}{2}, 8\frac{1}{2})$ $F(3\frac{1}{2}, \frac{1}{2}, 6\frac{1}{2})$. See Fig. 6-17.

Problem: Determine the true size of the dihedral angle between the two planes.

Solution:

Using the given data, lay out the plan and front elevation views of both planes. Draw inclined view 1 showing both planes and having the line *AB* in its true length. Draw inclined view 2 in order to show the line of intersection as a point and each of the planes as edges. The dihedral angle is measured in this view.

Ans. Dihedral Angle = 173°30'

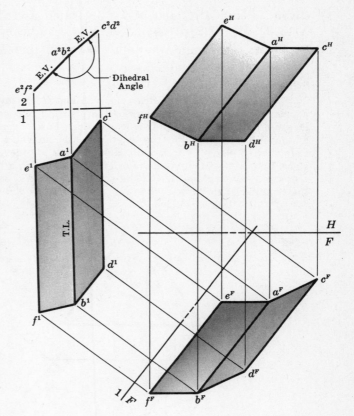

Fig. 6-17

12. Given: Fig. 6-18 shows the plan and front elevation views of three guy wires which are fastened to three mutually perpendicular planes. Scale: $1'' = 20'$.

Problem: Find the true length of each guy wire and the angle each wire makes with the plane to which it is attached.

Solution:

Draw auxiliary elevation view 1 to obtain the true length and slope of wire *AD*. Draw inclined view 2 in order to obtain the true length of *BD*. This view will also show the angle between wire *BD* and the back wall. Draw a profile view of *CD* and then project into the inclined view 3 which will show the true length of *CD* and the angle which wire *CD* makes with the side wall.

Ans. T.L. *AD* = 21', Slope *AD* = 54°30'
      T.L. *BD* = 14', Angle α = 58°30'
      T.L. *CD* = 21', Angle θ = 74°

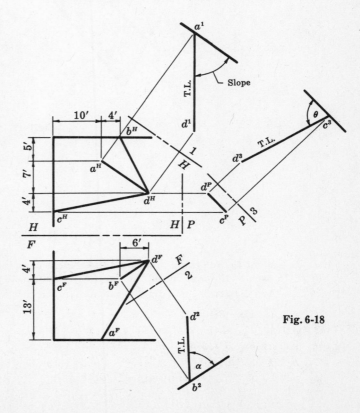

Fig. 6-18

13. Given: Plane ABC and line XY located as follows: $A(1,5,6)$ $B(2,3,7\frac{1}{2})$ $C(3,4\frac{1}{4},6)$ $X(1\frac{1}{2},3\frac{1}{2},6\frac{1}{4})$ $Y(2\frac{1}{4},4\frac{1}{4},6\frac{1}{2})$. See Fig. 6-19 below.

Problem: Determine the angle between line XY and the plane ABC. Use the complementary-angle method.

Solution:

Using the given data, draw the plan and front elevation views of plane ABC and line XY. Locate a true length line on both views of the plane. From point X in both views draw a line of indefinite length perpendicular to the true length line. For convenience stop the line at point Z in the front view so that YZ will be in its true length in the plan view. Draw auxiliary elevation view 1 to show the edge view of plane XYZ. Draw inclined view 2 in order to show the true size of plane XYZ. In view 2, since XZ is perpendicular to the original plane ABC, a line from X perpendicular to XZ will represent the edge view of plane ABC. The angle between XY and the line drawn perpendicular to XZ will be the required angle between line XY and the plane ABC.   *Ans.* Angle $= 52°30'$

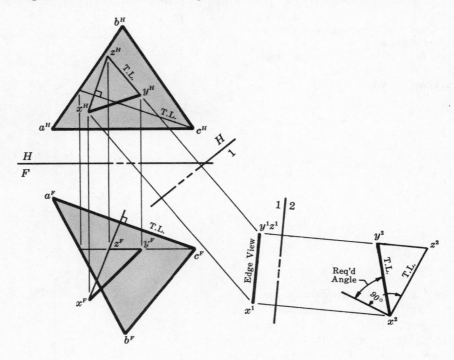

Fig. 6-19

Supplementary Problems

In each of the following problems 14 to 18 determine the dihedral angle formed by the planes ABC and BCD. In each of these problems the line BC is common to both planes.

See Art. 1.7 for the coordinate system of problem layout.

14. $A(1,1\frac{1}{2},4\frac{1}{2})$ $B(1,4,6\frac{1}{2})$ $C(3,3,5\frac{1}{2})$ $D(2\frac{1}{2},1\frac{1}{2},5)$ *Ans.* 18°30′

15. $A(1,2\frac{1}{2},3\frac{1}{2})$ $B(1\frac{1}{2},2,4)$ $C(2\frac{1}{2},\frac{1}{2},2\frac{1}{2})$ $D(2\frac{1}{2},2,4)$ *Ans.* 123°30′

16. $A(2\frac{1}{2},3,6)$ $B(\frac{1}{2},3,5\frac{1}{2})$ $C(1\frac{1}{2},1,4\frac{1}{4})$ $D(3,2,4\frac{1}{2})$ *Ans.* 27°30′

17. $A(1,\frac{1}{2},6)$ $B(3,3,6\frac{1}{2})$ $C(1,2\frac{1}{2},4)$ $D(3\frac{1}{2},4\frac{1}{4},5\frac{1}{4})$ *Ans.* 168°

18. $A(2\frac{1}{2},\frac{1}{2},6)$ $B(1\frac{1}{2},3,5)$ $C(4,\frac{1}{2},5)$ $D(2\frac{1}{2},\frac{1}{2},7)$ *Ans.* 19°

19. Two intersecting planes, *ABC* and *BCD*, are shown in
    Fig. 6-20. Scale: $\frac{1}{4}'' = 1'-0''$. Determine the dihedral
    angle between the two planes.   *Ans.* Angle = 100°

20. Two planes *ABD* and *ACD*. Point *B* is located 6′ east,
    4′ north of *A* and 6′ above *A*. Point *C* is 6′ east, 3′ south
    of *A* and 3′ above *A*. Point *D* is 10′ due east of *A* and
    3′ above *A*. Scale: $\frac{1}{4}'' = 1'-0''$. Determine the dihedral
    angle between the two planes.   *Ans.* Angle = 114°

21. Fig. 6-21 below shows a tetrahedron. Scale: 12″ = 1′-0″.
    Determine the angle the line *AB* makes with the plane
    ACD. Use only the edges necessary for solution.
    Ans. Angle = 51°

22. Two planes *ABC* and *BCD* are located as follows:
    A(3, 3, 7) *B*(4, 5, 7) *C*(4, 3, 6) *D*(5, 3, 7). Determine the
    dihedral angle between the two planes.
    Ans. Angle = 95°

23. The plan and front elevation views of a bridge pier are
    shown in Fig. 6-22 below. Scale: $\frac{3}{16}'' = 1'-0''$. Determine
    the dihedral angle between planes **A** and **B**.
    Ans. Angle = 95°

24. Scale: $\frac{3}{16}'' = 1'-0''$. Using the same figure as in Problem
    23 determine the dihedral angle between planes **B** and **C**.
    Ans. Angle = 100°

25. Fig. 6-23 below shows a hopper which feeds ore into open
    top freight cars. Scale: $\frac{1}{4}'' = 1'-0''$. Determine the
    dihedral angle between sides **A** and **B**.
    Ans. Angle = 99°

26. Using the same figure as in Prob. 25 determine the dihe-
    dral angle between planes *A* and *C*. Scale: $\frac{1}{4}'' = 1'-0''$.
    Ans. Angle = 27°30′

27. Planes *ABC* and *DEF*. Point *B* is 8′ east, 4′ north of *A*
    and 9′ above *A*. Point *C* is 12′ east, 2′ south of *A* and
    4′ above *A*. Point *D* is 2′ east, 4′ north of *A* and 8′ above
    A. Point *E* is 10′ east, 4′ south of *A* and 1′ above *A*.
    Point *F* is 14′ east, 5′ north of *A* and 4′ above *A*. Scale:
    $\frac{1}{4}'' = 1'-0''$. Determine the dihedral angle between the
    two planes.   *Ans.* Angle = 44°

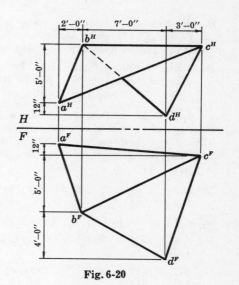

Fig. 6-20

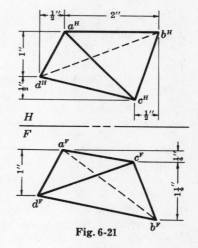

Fig. 6-21

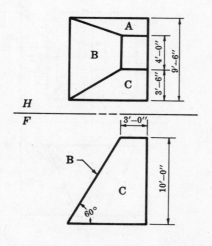

Fig. 6-22

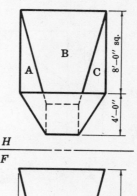

Fig. 6-23

28. A steel plate transition piece is shown in Fig. 6-24. Scale: $\frac{1}{4}'' = 1'-0''$. Determine the dihedral angle between planes A and B.   *Ans.* Angle = 80°

29. Using the same figure as in Problem 28 determine the dihedral angle between planes B and C. Scale: $\frac{1}{4}'' = 1'-0''$.   *Ans.* Angle = 106°

In each of the following problems 30 to 35 determine the true size of the angle between the line *XY* and the plane *ABC*. Use either the line method or the plane method.

    See Art. 1.7 for the coordinate system of problem layout.

30. $A(1, 1\frac{1}{2}, 5)$ $B(2, 2\frac{1}{2}, 4)$ $C(3\frac{1}{2}, \frac{1}{2}, 5\frac{1}{2})$ $X(1\frac{1}{2}, \frac{1}{2}, 4\frac{1}{4})$ $Y(3, 2, 4\frac{1}{2})$

31. $A(1\frac{1}{4}, 2, 3\frac{1}{2})$ $B(2\frac{1}{4}, 1, 5\frac{1}{4})$ $C(4, 1\frac{1}{2}, 3\frac{1}{2})$ $X(1\frac{1}{2}, 2\frac{1}{4}, 5)$ $Y(3\frac{1}{4}, 1\frac{1}{2}, 3\frac{1}{4})$

32. $A(1\frac{1}{2}, 1\frac{1}{2}, 5)$ $B(2, 3\frac{1}{4}, 4\frac{1}{2})$ $C(3\frac{1}{2}, 1, 6)$ $X(1, 2, 5)$ $Y(3\frac{1}{2}, 2\frac{1}{4}, 4)$

33. $A(\frac{1}{4}, 3\frac{1}{2}, 5\frac{1}{2})$ $B(1\frac{1}{2}, 3\frac{1}{2}, 7)$ $C(1\frac{1}{4}, 1\frac{1}{2}, 4\frac{1}{2})$ $X(2, 2\frac{1}{4}, 4\frac{1}{2})$ $Y(1\frac{1}{2}, 3, 6)$

34. $A(1, 3\frac{1}{2}, 6\frac{1}{2})$ $B(2\frac{1}{4}, 2\frac{1}{2}, 5\frac{1}{4})$ $C(3\frac{1}{2}, 4\frac{1}{4}, 6\frac{1}{4})$ $X(1\frac{1}{2}, 3\frac{1}{4}, 5\frac{1}{4})$ $Y(3, 3\frac{1}{4}, 6\frac{1}{2})$

35. $A(1\frac{1}{2}, 1\frac{1}{2}, 6\frac{1}{2})$ $B(2, 3, 5\frac{1}{4})$ $C(4, 2, 5)$ $X(1, 2, 5)$ $Y(3, 2, 5\frac{1}{4})$

    Ans. 30. 38°30'   32. 37°30'   34. 43°
            31. 25°   33. 22°30'   35. 26°30'

36. Fig. 6-25 shows a pyramid intersected by a line *XY*. Scale: $12'' = 1'-0''$. Determine the angle between the line and the plane *OCD*.   *Ans.* Angle = 69°30'

37. Line *XY* represents one of several brace rods which support two intersecting billboards as shown in Fig. 6-26 below. Scale: $\frac{1}{8}'' = 1'-0''$. Determine the angle that the rod makes with the billboard at anchor *Y*.   *Ans.* Angle = 36°

38. Using the same data as in Problem 37 determine the angle that the rod makes with the billboard at anchor *X*. Scale: $\frac{1}{8}'' = 1'-0''$.   *Ans.* Angle = 51°30'

39. Fig. 6-27 below shows the plan and front elevation views of a hopper which is fed by a pipe represented by the centerline *XY*. Scale: $\frac{3}{8}'' = 1'-0''$. Determine the angle that the pipe makes with the wall.   *Ans.* Angle = 37°30'

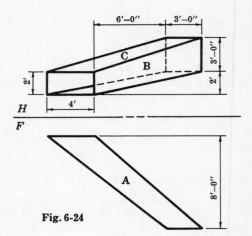

Fig. 6-24

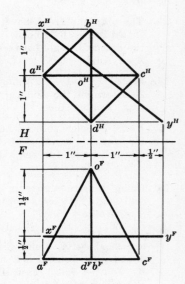

Fig. 6-25

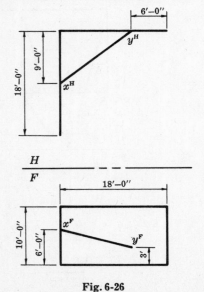

Fig. 6-26

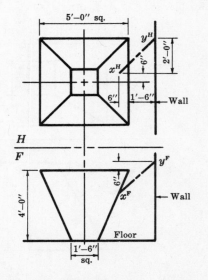

Fig. 6-27

40. Fig. 6-28 below shows a television antenna supported by two brace rods attached to the roof of a house. Scale: $\frac{1}{8}'' = 1'-0''$. What angles do the braces make with the roof?

Ans. $XY = 70°$ $XZ = 48°$

41. Fig. 6-29 below shows a regular hexagonal pyramid. Scale: $12'' = 1'-0''$. Determine the dihedral angle between planes AOB and AOF.   *Ans.* Angle $= 125°$

42. Using the same figure as for problem 41 determine the angle between line OC and plane AOF. Scale: $12'' = 1'-0''$.   *Ans.* Angle $= 45°$

43. There is possible interference between a pipe XY and a plate $ABCD$. Determine whether or not there is interference and find the angle between the pipe and the plate. $X(1\frac{1}{2}, 5, 7\frac{1}{2})$ $Y(3, 5\frac{1}{2}, 7\frac{1}{2})$ $A(2, 6, 7)$ $B(3, 4\frac{1}{2}, 8)$ $C(4\frac{1}{2}, 4\frac{1}{2}, 7\frac{1}{2})$ $D(3\frac{1}{2}, 6, 6\frac{1}{2})$. Show the pierce point of the line and plane in all views. Also show proper visibility. See Art. 1.7 for the coordinate system of problem layout.

Ans. No interference. Angle $= 26°$

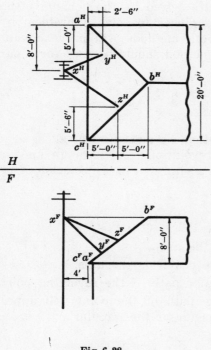

Fig. 6-28

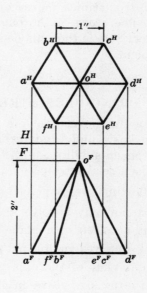

Fig. 6-29

Chapter 7

Revolution

The method of solving Descriptive Geometry problems, as already discussed in preceding chapters, is based on what is called "change-of-position". The various views which were obtained depended upon the fact that the observer assumed a different position of viewing the stationary point, line, or plane in space. However, many basic problems can be solved by having the observer remain stationary while the object is revolved to a position which will reveal certain required information.

Even though the majority of space problems are solved by the "change-of-position" method, it is sometimes more convenient for the engineer to solve a problem by using revolution. Therefore, a thorough knowledge of both methods will prove advantageous to the engineer. It is then his decision as to which method should be used for a particular type of problem.

7.1 BASIC PRINCIPLES of REVOLUTION

Before a student can solve a problem by revolution, he must first of all be acquainted with certain basic and fundamental principles of revolution. A thorough knowledge of these basic concepts will make the revolution method a vital tool in the solution of problems involving space relationship.

The following are principles of revolution:

(1) A point can only revolve about a straight line axis.
(2) A point revolves in a plane perpendicular to the axis.
(3) A point revolves in the path of a circle whose center is the axis.
(4) The point view of the axis will show the circular path of the revolving point.
(5) When the axis appears in its true length, the path of the point will appear as a straight line whose length is equal to the diameter of the circular path.

The student should study Fig. 7-1 to make absolutely certain that he thoroughly understands the underlying principles of revolution. Of course, failure to understand these basic concepts may result in assumptions which are contrary to fact.

In many Descriptive Geometry problems, the axis about which a point revolves does not appear in its true length in either the plan or front elevation views. Therefore, a new view must be drawn which will show the true length of the axis, then an additional view will show the point view of the axis.

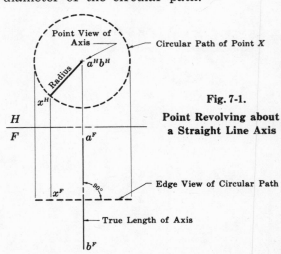

Fig. 7-1.

Point Revolving about a Straight Line Axis

In Fig. 7-2, the oblique line axis is shown in both the plan and front elevation views. Point X which revolves about this axis is also located in the two given views. Using the "change-of-position" method, a new view is drawn which shows the axis, AB, in its true length and the point X located at a fixed distance from the axis. Inclined view 2 determines the point view of the axis and the circular path of point X as it revolves about the axis.

Should it be required to locate the lowest and highest positions of point X as it revolves about the oblique line axis, the extreme points of the circular path shown as an edge in view 1 will determine these positions. In view 2 the low and high positions of point X are designated by the additional subscript letters L and H.

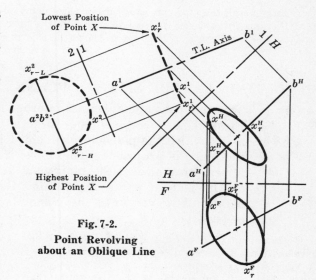

Fig. 7-2.
Point Revolving about an Oblique Line

7.2 To FIND the TRUE LENGTH of a LINE

Analysis: If a line is parallel to a projection plane, it will be projected on that plane in its true length. Therefore, if a line in space is revolved about an axis until it lies parallel to a projection plane, it will then appear in its true length.

Example: In Fig. 7-3, the oblique line AB appears in both the plan and front elevation views. Using a vertical axis through point A in the plan view, point B should be revolved until line AB lies parallel to the frontal plane. It will then appear in its true length in the front view. Note that the revolving of point B does not alter its elevation. The true slope of the line can also be seen in the front view since the line appears in its true length in an elevation view.

Fig. 7-4 below shows the same oblique line being revolved about a horizontal axis. At (a) the axis passes through one end of the line and, at (b) the axis passes through the mid-point of the line.

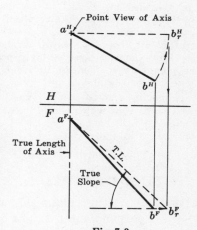

Fig. 7-3.
True Length of a Line by Revolving about a Vertical Axis

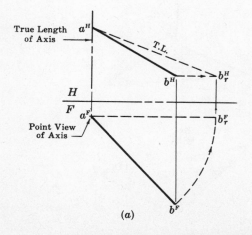

(a)

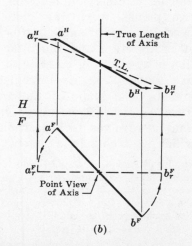

(b)

Fig. 7-4.   True Length of a Line by Revolving about a Horizontal Axis

In general, the axis can pass through any point of the line. For convenience, it is usually better to pass the axis through one of the end points of the line. In this case only one point needs to be revolved to show the true length of the line as shown in Figures 7-3 and 7-4(a).

7.3 To FIND the TRUE SIZE of a PLANE

Analysis: Two conditions must be met in order for a plane to be revolved into its true size. First, it must revolve about an axis which lies on the plane. Second, the axis about which the plane revolves must lie parallel to the image plane upon which the plane is projected. Therefore, if the plane is to be revolved so that it will appear true size in the plan view, it must be revolved about a level line axis on the plane. After determining the location of the axis, each point on the plane is revolved until the plane is level.

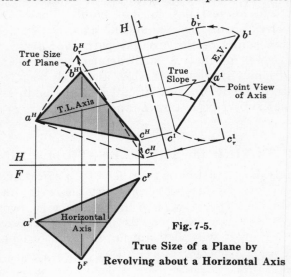

Fig. 7-5.

True Size of a Plane by Revolving about a Horizontal Axis

Example: In Fig. 7-5, the plane ABC is to be revolved until it lies parallel to the horizontal plane. A horizontal axis is drawn in the front elevation view which will appear in its true length in the plan view and as a point in the auxiliary elevation view 1. The plane is revolved about the point view of the axis in view 1 until it lies parallel to the horizontal plane. Project the revolved points B and C to the plan view until they intersect the projections of points B and C drawn parallel to the H-1 folding line. Connect these revolved points B and C in the plan view to point A in order to determine the true size of the plane.

It should be noted that a plane will appear as an edge in the view which shows the axis as a point. This edge view, being an elevation view, also shows the true slope of the plane.

7.4 To FIND the DIHEDRAL ANGLE

Analysis: A cutting plane which is passed perpendicular to the intersection of two planes determines the dihedral angle between the two planes. After determining the line of intersection between the two planes, a cutting plane is passed at a right angle to this line of intersection and the intersection of the cutting plane with the two given planes is thus determined. The plane of the angle formed by these two intersection lines is revolved to a position in which its true size is shown.

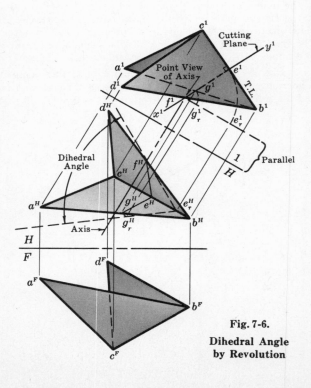

Fig. 7-6.

Dihedral Angle by Revolution

Example: In Fig. 7-6, the plan and front elevation views of planes ABC and BCD are given. An auxiliary elevation view is drawn showing both planes with their common line of intersection BC appearing in its true length. A cutting plane XY is drawn perpendicular to line BC at any point, in this case E. The cutting plane cuts the two given planes at F, E, and G. Project these points back to the plan view. These points determine the dihedral angle which now must be revolved about a level-line axis through a point such as F. The axis appears as a point in view 1. Revolve points G and E through the horizontal axis at F until they lie in a plane parallel to the horizontal image plane. The dihedral angle now revolved to a horizontal plane will appear in its true size in the plan view.

7.5   To FIND the ANGLE BETWEEN a LINE and a PLANE

Analysis: In order to determine the angle between a line and a plane by revolution, the line must be revolved about an axis which is perpendicular to the plane. Revolution about any other axis will not show the true angle desired. A view showing the true size of the plane will also show the axis as a point. Therefore the axis will appear in its true length in the view showing the plane as an edge and will be perpendicular to this edge view of the plane. The line is revolved until it appears in its true length in the view showing the plane as an edge. The true angle between the line and the plane is measured in this view.

   Example: In Fig. 7-7 below, the plane ABC and the line XY are given in both the plan and front elevation views. An auxiliary elevation view 1 is drawn showing the line XY and the edge view of the plane ABC. An inclined view is then drawn which would show the true size of the plane, but drawing the true size of the plane in this view will serve no useful purpose since we are only interested in showing the point view of the axis about which the line XY will revolve. The line XY is revolved until it appears in its true length in the auxiliary elevation view 1 where the angle can be measured.

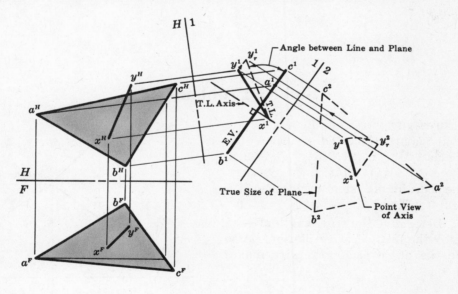

Fig. 7-7.   Angle between a Line and a Plane by Revolution

7.6   To FIND a LINE at GIVEN ANGLES with TWO PRINCIPAL PLANES

Analysis: A right circular cone will have elements having the same length, and these elements will make the same angle with the base of the cone. The intersection of two cones having elements the same length, with a common vertex, and with axes perpendicular to each other, will determine the required line. The two cones must intersect along a common element line making the required angles with both principal planes which are perpendicular to each other.

Depending upon the location of the cone bases, there are eight possible solutions to the problem. Since the two principal planes are perpendicular to each other, the sum of the two given angles cannot be more than 90°. If the sum of the two given angles is exactly 90°, then the two cones are obviously tangent.

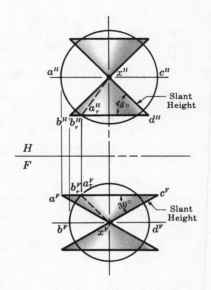

Example: It is required to locate a line which makes a 45° angle with the frontal plane and a 30° angle with the horizontal plane (see Fig. 7-8).

Determine a convenient location for point X in both the plan and front elevation views. This point will be the common vertex of two right circular cones having the same slant height. In the front view, construct the right circular cone XAC having a given slant height and making an angle of 30° with the horizontal base plane which is, of course, parallel to the horizontal projection plane. Show this cone in the plan view. In the plan view, construct the right circular cone XBD having the same slant height as cone XAC and making an angle of 45° with the vertical base plane which is parallel to the frontal projection plane. Show this cone in the front view. The intersection of the revolved positions of points A and B will determine the required line from vertex X. Note the possible alternate positions of the cones.

Fig. 7-8.

Locating a Line Making Given Angles with Two Principal Planes

Solved Problems

1. **Given:** Three guy wires, A, B, and C, are attached at point O located 2′ below the top of a mast, XY. $A(2\frac{1}{2}, 3, 5\frac{1}{2})$ $B(3\frac{1}{2}, 2, 8)$ $C(6, 2\frac{1}{2}, 6)$ $X(4\frac{1}{2}, 4\frac{1}{2}, 7)$ $Y(4\frac{1}{2}, 2, 7)$. See Art. 1.7 for the coordinate system of problem layout. Refer to Fig. 7-9. Scale: $\frac{1}{8}'' = 1'-0''$.

Problem: Determine the true length of each guy wire by the method of revolution. Also determine the angle each guy wire makes with the mast.

Solution:

Using the given data, draw the guy wires and mast in both the plan and front elevation views. In the plan view, using the mast as a vertical axis, revolve points A, B, and C until they lie in a plane parallel to the frontal plane. From points A, B, and C in the front view, draw level lines until they intersect the projections of A, B, C, revolved. Connect each of these intersecting points to point O in the front view. Both the true length of each guy wire and the angle it makes with the mast can be measured in this front elevation view.

Ans. Guy wire AO: T.L. = 10′-9$\frac{1}{2}$″   Angle = 68°
       Guy wire BO: T.L. = 9′-10″   Angle = 35°
       Guy wire CO: T.L. = 9′-5″   Angle = 50°

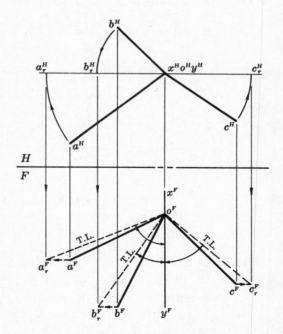

Fig. 7-9

2. **Given:** Plane *ABCD*. Point *B* is located 1″ east, 1½″ north of *A* and ½″ below *A*. Point *C* is located 3″ east, 2″ north of *A* and 1½″ below *A*. Point *D* is located 2″ due east of *A* and 1″ below *A*. See Fig. 7-10. Scale: ½″ = 1″.

Problem: By the method of revolution, show the true size of the plane in the plan view.

Solution:

Draw the plan and front elevation views of plane *ABCD*. Since the front view shows the edge view of the plane, a level axis is assumed through point *A* in the front view and points *B*, *C*, and *D* are then revolved until they lie in a horizontal plane. From points *B*, *C*, and *D* in the plan view, frontal lines are drawn until they intersect the projections of *B*, *C*, and *D* revolved. Connect the three points of intersection as shown to obtain the true size of the plane.

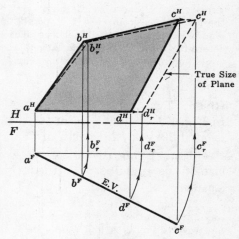

Fig. 7-10

3. **Given:** Plane *ABC*. Point *B* is located 5′ east, 6′ north of *A* and 6′ below *A*. Point *C* is located 10′ east, 5′ north of *A* and 4′ below *A*. See Fig. 7-11. Scale: ⅛″ = 1′-0″.

Problem: Determine the true size of each plane angle by the methods of revolution.

Solution:

Using the given data, construct the plan and front elevation views of plane *ABC*. Draw auxiliary elevation view 1 to show the plane as an edge. Assume a level axis through point *A* in view 1 and revolve points *B* and *C* until they lie in a horizontal plane. Project points *B* and *C* into the plan view until they intersect the projections of *B* and *C* drawn parallel to the *H*-1 folding line in the plan view. These intersections determine the positions of points *B* and *C* revolved, thus showing the true size of the plane. Each of the plane angles can be measured in this view.

Ans. Angle $A = 28°$, Angle $B = 97°$, Angle $C = 55°$

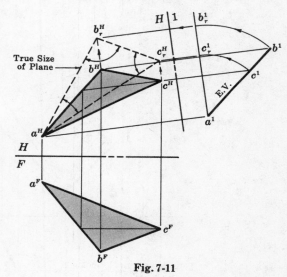

Fig. 7-11

4. **Given:** Plane *ABC* is located as follows: Point *B* is 5′ east, 4′ north of *A* and 5′ above *A*. Point *C* is located 9′ east, 4′ south of *A* and 2′ above *A*. See Fig. 7-12. Scale: ⅛″ = 1′-0″.

Problem: Using the methods of revolution, determine the diameter of the largest circle to be drawn within the limited plane, *ABC*.

Solution:

Using the given data, draw the plan and front elevation views of plane *ABC*. Draw auxiliary elevation view 1 to show the plane as an edge. Assume a level axis through point *A* and revolve points *B* and *C* in view 1 until they lie in a horizontal plane.

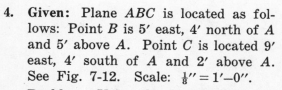

Fig. 7-12

From points B and C in the plan view, draw lines parallel to the H-1 folding line until they intersect the projections from view 1 of points B and C revolved. Connect these points of intersection to the plan view of point A. The true size of the plane ABC is now established. Angle bisectors will determine the center for the largest circle to be inscribed within the limited plane ABC. Draw the circle and measure the diameter using the given scale.   *Ans.* Diameter $= 5'-2''$

5. **Given:** Two water lines from points A and B converge at a common point C as shown in Fig. 7-13 below. Point A is 30' west, 30' north of C and 30' above C. Point B is located 20' east, 20' north of C and 40' above C. Scale: $1'' = 30'$.

    Problem: Using revolution, determine the true length of each water line and the true size of the angle between them.

    Solution:

    Using the given data, draw the plan and front elevation views of both water lines. Draw auxiliary elevation view 1 to show the edge view of plane ABC. Assume a level axis through point C and revolve points A and B in view 1 until they lie in a horizontal plane. From points A and B in the plan view, draw lines parallel to the H-1 folding line until they intersect the projections from view 1 of points A and B revolved. Connect these points of intersection to point C in the plan view and measure the required angle. The plan view will also show each water line in its true length.

    Ans. T.L. $AC = 52'-0''$, T.L. $BC = 49'-4''$, Angle $ACB = 61°30'$

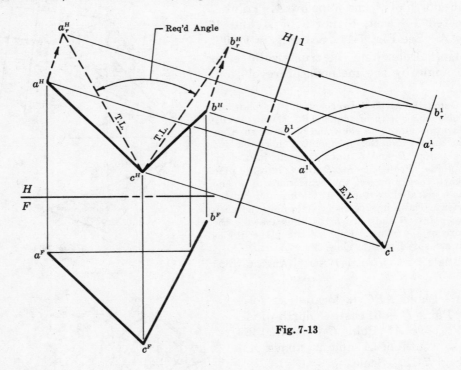

Fig. 7-13

6. **Given:** Two support braces intersect at a common point C as shown in Fig. 7-14 below. Point A is located 6' west, 2' south of C and at the same elevation as C. Point B is located 3' east, 4' south of C and 3' above C. Scale: $\frac{3}{16} = 1'-0''$.

    Problem: By the methods of revolution, determine the angle between the two braces. Check your answer by an independent method.

    Solution:

    Draw the plan and front elevation views of the two braces AC and BC. Draw auxiliary elevation view 1 to show the plane ABC as an edge. Assume a level axis through point B in view 1 and revolve both A and C until they lie in a horizontal plane parallel to H-1. From points A and C in the plan view, draw lines parallel to the H-1 folding line until they intersect the projections of points A and C

revolved. Connect the revolved positions of A and C in the plan view. Also connect point B to the plan view of C revolved. The required angle can now be measured in the plan view.

The answer can be checked by drawing inclined view 2 which will show the true size of plane ABC. Angle C is the required angle in this view.   *Ans.* 105°30′

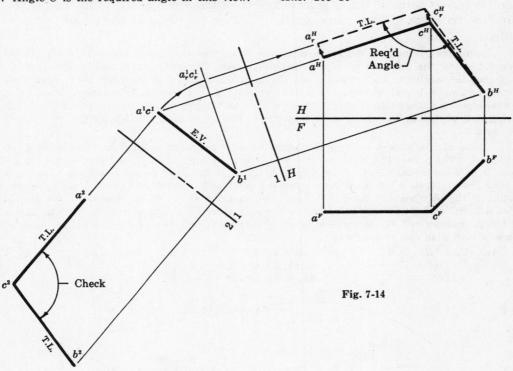

Fig. 7-14

7. Given: Lines AB and BC in Fig. 7-15 represent two supporting braces of an engine mount. Point B is 9″ east, 1′ north of A and 9″ below A. Point C is 1′−10″ due east of A and 7″ above A. Scale: $\frac{3}{4}″ = 1′−0″$.

Problem: Using the revolution method, determine the true length of each supporting brace and the true size of the angle between them.

Solution:

Using the given data, draw the plan and front elevation views of the two braces. Draw auxiliary elevation view 1 which will show the edge view of plane ABC. Assume a level axis through point A and revolve points B and C until they lie in a horizontal plane. From points B and C in the plan view, draw lines parallel to the H-1 folding line until they intersect the projections of points B and C revolved. Connect these intersecting points from B and C revolved and from point A to B revolved in the plan view. This view yields the true length of each brace as well as the true size of the angle between them.

Ans. T.L. $AB = 1′-5\frac{1}{2}″$, T.L. $BC = 2′-0″$,
Angle $ABC = 66°$

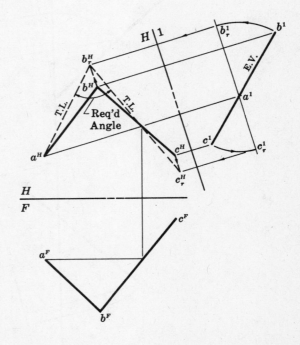

Fig. 7-15

8. **Given:** Plane *ABC* shown in Fig. 7-16 below. Point *B* is 1′ west, 3′ north of *A* and 4′ below *A*. Point *C* is 7′ east, 5′ north of *A* and 9′ below *A*. Scale: $\frac{3}{16}'' = 1'-0''$.

   Problem: Revolve the plane until it appears true size in both the plan and front elevation views. Determine the true length of the three lines which limit the plane. Determine the true size of each plane angle. What is the slope of the plane?

   Solution:

   Using the given data, draw the plan and front elevation views of plane *ABC*. Draw inclined view 1 to show the plane as an edge. Assume an axis through point *B* and revolve points *A* and *C* until they lie parallel to the frontal plane. From points *A* and *C* in the front view, draw lines parallel to folding line *F*-1 until they intersect the projections from points *A* and *C* revolved. Connect the revolved points to point *B* in the front view to obtain the true size of the plane. The true size of each plane angle can be measured in this view. The front view also shows the true length of *AB*, *AC*, and *BC*.

   Draw auxiliary elevation view 2 in order to show the plane as an edge. The slope of the plane is measured in this view. In view 2, assume a level axis through point *A*, and revolve both *B* and *C* until they lie in a horizontal plane. From points *B* and *C* in the plan view, draw lines parallel to the *H*-2 folding line until they intersect the projections of *B* and *C* revolved. Connect these points of intersection to the plan view of point *A*. Plane *ABC* is again shown in its true size. The true lengths of lines *AB*, *BC*, and *AC* can be measured in the plan view. The true size of each plane angle can also be found in this view. If the problem is solved accurately, the size of plane *ABC* revolved will be the same in both plan and front views.

   Ans. T.L. $AB = 5'-1''$ Angle $A = 46°$ Slope $= 56°$
   T.L. $AC = 12'-6''$ Angle $B = 111°30'$
   T.L. $BC = 9'-7\frac{1}{2}''$ Angle $C = 22°30'$

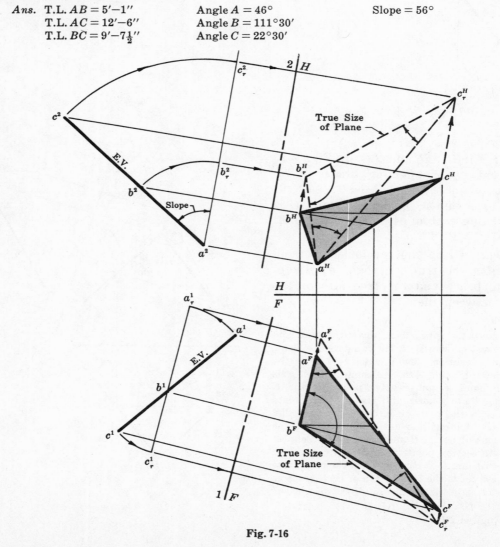

Fig. 7-16

9. Given: Plan and front elevation views of a grain hopper. Refer to Fig. 7-17. Scale: $\frac{1}{8}'' = 1'-0''$.

Problem: Determine the dihedral angle between planes *ABCD* and *BDEF*.

Solution:

Draw an auxiliary elevation view showing the line of intersection, *BD*, in its true length. Pass a cutting plane perpendicular to *BD* at any point, such as *Y*. The dihedral angle is the plane angle between *YX* and *YZ*. To obtain the true size of the dihedral angle in the plan view, the cutting plane must be revolved about a level line axis such as *XZ*. In the elevation view 1, revolve point *Y* to a level position and project back to the plan view in order to obtain the true size of the dihedral angle.

Ans. Angle $= 97°30'$

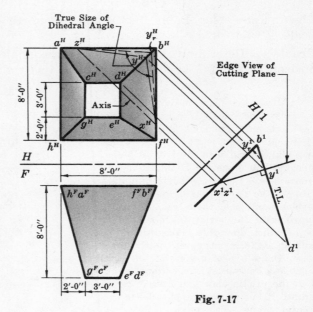

Fig. 7-17

10. Given: Plane *ABC* and line *BX* in Fig. 7-18 below. Point *B* is 6′ east, 10′ north of *A* and 4′ below *A*. Point *C* is 17′ east, 2′ south of *A* and 9′ above *A*. Point *X* is 20′ east, 8′ north of *A* and 3′ below *A*. Scale: $\frac{1}{16}'' = 1'-0''$.

Problem: Using the revolution method, determine the angle between line *BX* and the plane *ABC*. What is the true size of angle *ACB*?

Solution:

Use the given data to lay out the plan and front elevation views of plane *ABC* and line *BX*. Draw auxiliary elevation view 1 to show the line *BX* and the edge view of plane *ABC*. Construct inclined view 2 in order to show line *BX* and the true size of plane *ABC*. In view 2, revolve point *X* about the axis through *B* until it lies parallel to folding line 1-2. In elevation view 1, draw a line from point *X* parallel to folding line 1-2 until it intersects the projection of point *X* revolved. From this intersecting point, draw a line to point *B*. The angle between the line and plane can now be measured in view 1. Plane angle *ACB* is measured in view 2.

Ans. Angle $= 20°30'$

Angle $ACB = 36°$

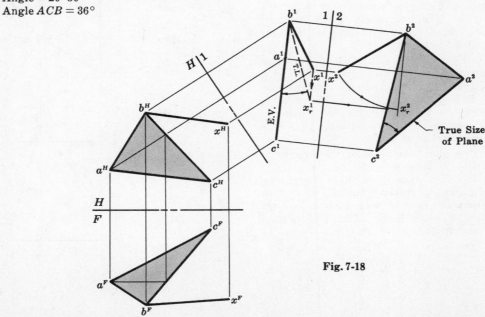

Fig. 7-18

11. Given: Line AB. Point B is 8′ east, 4′ north of A and 7′ below A. See Fig. 7-19. Scale: $\frac{1}{8}'' = 1'-0''$.

Problem: Determine the true size of the angles which the line AB makes with the horizontal, frontal, and profile planes. Use the revolution method.

Solution:

Draw the plan and front elevation views of line AB. Assume a vertical axis through point B in the plan view and revolve point A until it lies in a plane parallel to the frontal plane. Measure the slope, or H angle, in the front view. Assume a horizontal axis through the front view of point A and revolve point B until it lies in a plane parallel to the horizontal plane. The angle which line AB makes with the frontal plane is measured in the plan view. Draw a profile view of line AB. Assume a horizontal axis through point B in the profile view and revolve point A until it lies in a plane parallel to the frontal plane. The angle which line AB makes with the profile view is measured in the frontal view.

Ans. $H = 38°$, $F = 20°30'$, $P = 45°$

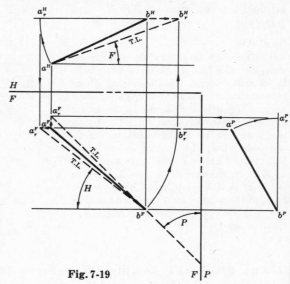

Fig. 7-19

12. Given: Fig. 7-20 shows a flagpole, XY, supported by three guy wires: XA, XB, and XC. Scale: $\frac{1}{16}'' = 1'-0''$.

Problem: Determine, by revolution, the true length of each guy wire and the angle it makes with the surface to which it is anchored.

Solution:

Draw the plan and front elevation views of the flagpole, guy wires, and three mutually perpendicular surfaces. Draw a profile view of wire XA. Assume a level axis through point X in this view and revolve point A until it lies in a horizontal plane. Project the revolved position of A into the plan view and connect to point X in this view. The true length of the guy wire and the angle it makes with the side wall can be measured in this view.

In the front view, assume a level axis through point X and revolve point B until it lies in a horizontal plane. Project the revolved position of B into the plan view and connect to point X in the plan view. The true length of XB and the angle it makes with the back wall can be measured in this view.

In the plan view, assume the axis to be the vertical flagpole and revolve the point C until it lies in a plane parallel to the frontal plane. Project the revolved position of C down to the front view and connect to point X in the front view. The true length of XC and its slope can be measured in this view.

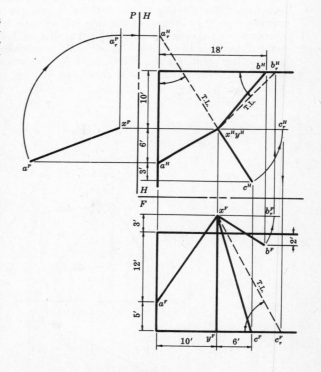

Ans.

Guy Wire	T.L.	Angle
XA	19′–0″	32°
XB	13′–10″	46°30′
XC	22′–10″	61°30′

Fig. 7-20

13. Given: Plane *ABC* and line *XY* in Fig. 7-21 below. Point *B* is 7′ due east of *A* and 1′ above *A*. Point *C* is 4′ east, 8′ south of *A* and 6′ above *A*. Point *X* is 9′ east, 1′ south of *A* and 5′ above *A*. Point *Y* is 4′ east, 6′ south of *A* and 4′ above *A*. Scale: $\frac{1}{8}'' = 1'-0''$.

Problem: Using the revolution method, determine (*a*) the true size of the angle between line *XY* and the plane *ABC*, (*b*) angle *BAC*, (*c*) the true length of line *XY*.

Solution:

Using the given data, draw the plan and front elevation views of the line and plane. Draw inclined view 1 to show the line *XY* and the edge view of plane *ABC*. Construct inclined view 2 which will show the line *XY* and the true size of plane *ABC*. In view 2, revolve point *X* about the axis through *Y* until *X* is lying on a plane parallel to folding line 1-2. From point *X* in view 1, draw a line parallel to folding line 1-2 until it intersects the projection of point *X* revolved. Connect point *X* revolved in view 1 to point *Y* in this view. This line is the true length of *XY*. Measure the angle between the true length of *XY* and the edge view of the plane. Angle *BAC* would be measured in view 2.

Ans. (*a*) Angle = 25°, (*b*) Angle *BAC* = 63°, (*c*) T.L. *XY* = $7'-2\frac{1}{2}''$

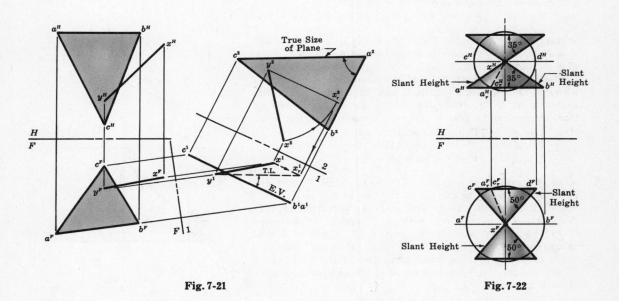

Fig. 7-21   **Fig. 7-22**

14. Given: Plan and front elevation views of vertex point *X* and slant height of 1″ for two intersecting cones. See Fig. 7-22 above. Scale: $\frac{1}{2}'' = 1''$.

Problem: Locate a line through point *X* which shall make an angle of 35° with the frontal plane and an angle of 50° with the horizontal plane.

Solution:

Locate point *X* in both plan and front views. This point will be the common vertex of two right circular cones both having a slant height of 1″. In the plan view construct the right circular cone *XAB* having a slant height of 1″ and making an angle of 35° with the vertical base plane which is parallel to the frontal plane. Show this cone in the front view. In the front elevation view construct the right circular cone *XCD* having a slant height of 1″ and making an angle of 50° with the horizontal base plane. Show this cone in the plan view. The intersection of the revolved positions of points *A* and *C* will determine the line which makes an angle of 35° with the frontal plane and an angle of 50° with the horizontal plane. Fig. 7-22 shows alternate positions of cones.

Supplementary Problems

In the following problems determine, by revolution, the true length and slope of line AB:
See Art. 1.7 for coordinate system of problem layout.

				T.L.	Slope
15.	$A(1, 1, 5\frac{1}{2})$ $B(3, 3, 4\frac{1}{2})$	Scale: $1'' = 1'-0''$	Ans.	$3'-0''$	$41°30'$
16.	$A(2, 4, 6)$ $B(4, 1, 5)$	Scale: $\frac{1}{4}'' = 1'-0''$	Ans.	$15'-0''$	$53°$
17.	$A(1\frac{1}{2}, 2\frac{1}{2}, 5)$ $B(3\frac{1}{2}, 4, 6)$	Scale: $1'' = 20'$	Ans.	$53'-9''$	$34°$
18.	$A(2\frac{1}{2}, 1, 4\frac{1}{2})$ $B(5, 1, 5\frac{1}{2})$	Scale: $\frac{1}{2}'' = 1'-0''$	Ans.	$5'-4\frac{1}{2}''$	$0°$
19.	$A(\frac{1}{2}, 4, 6\frac{1}{2})$ $B(4\frac{1}{2}, 2, 5\frac{1}{4})$	Scale: $1'' = 10'$	Ans.	$46'-6''$	$25°30'$

20. A short-wave radio antenna, AB, is supported by three guy wires located as follows: $A(5, 5, 7\frac{1}{2})$ $B(5, 0, 7\frac{1}{2})$ $C(5, 1, 5)$ $D(1\frac{1}{2}, 2\frac{1}{2}, 8)$ $E(7, 1\frac{1}{2}, 9)$. Scale: $\frac{1}{4}'' = 1'-0''$. Determine the true length of each guy wire by using the revolution method.

   Ans. T.L. $AC = 18'-11''$, T.L. $AD = 18'-0''$, T.L. $AE = 17'-3''$

21. A flagpole AB, shown in Fig. 7-23, is 12' high. Its stability is maintained by three guy wires located as follows: Point C is 8' south, 7' east of AB and 2' higher than B. Point D is 4' south, 9' west of AB and is fastened to an anchor 3' above B. Point E is 8' north, 2' east of B and at the same elevation as B. Scale: $\frac{3}{16}'' = 1'-0''$. By means of revolution, determine the true length and slope of each guy wire.

   Ans. T.L. $AC = 14'-6''$ Slope $AC = 43°$
         T.L. $AD = 13'-4''$ Slope $AD = 42°$
         T.L. $AE = 14'-6''$ Slope $AE = 56°$

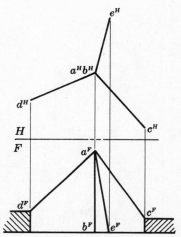

22. A short-wave radio antenna, XY, is 15' high. Three guy wires are attached to the top of the antenna. See Fig. 7-24 below. Point Y on the ground is at an elevation of 137'. Point A is 13' west, 3'-6'' north of X and at an elevation of 141'-6''. Point B is 5'-6'' east, 9'-6'' south of X and at an elevation of 139'-6''. Point C is 12'-6'' east, 10' north of X and at the same elevation as B. Scale: $\frac{1}{8}'' = 1'-0''$. Determine, by revolution, the true length and slope of each guy wire.

   Ans. T.L. $AX = 17'-0''$ Slope $AX = 38°$
         T.L. $BX = 16'-8''$ Slope $BX = 48°30'$
         T.L. $CX = 20'-4''$ Slope $CX = 38°$

Fig. 7-23

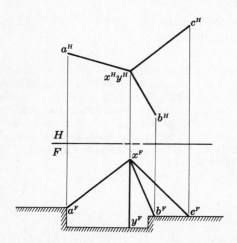

Fig. 7-24

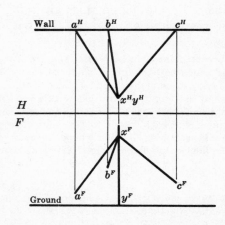

Fig. 7-25

23. A flagpole is held rigid by three pipe braces attached to a vertical wall as shown in Fig. 7-25 above. Point B is 5′–6″ east of A and 4′–6″ above A. Point C is 17′ east of A and 2′ above A. Point X is 7′–6″ east, 10′ above A and located 12′ from the wall. Point Y is 7′–6″ east, 12′ south of A and 2′ below A. Scale: $\frac{1}{8}″ = 1′–0″$. By means of revolution, determine the true length of each pipe brace. *Ans.* T.L. $AX = 17′–4″$, T.L. $BX = 13′–4″$, T.L. $CX = 17′–4″$

24. Using the same data and scale as used in Problem 23, determine, by the revolution method, the angle each pipe brace makes with the vertical wall.   *Ans.* Angle $A = 44°$, Angle $B = 64°$, Angle $C = 44°$

25. By the revolution method, determine the true size of each plane angle in the plane ABC. $A(1, 1\frac{1}{2}, 8)$ $B(4, 4, 8\frac{1}{2})$ $C(4, 2, 6)$.   *Ans.* Angle $A = 50°$, Angle $B = 60°$, Angle $C = 70°$

26. Line AB and point X are located as follows: $A(4, 1\frac{1}{4}, 8)$, $B(6, 4\frac{1}{2}, 6)$, $X(5\frac{1}{4}, 5, 7\frac{1}{2})$. Scale: $\frac{1}{4}″ = 1′–0″$. As point X revolves about the axis AB, what is the diameter of the moving point?   *Ans.* 12′–10″

27. Two sewer lines from points A and B converge at a manhole, C. Point B is 10′ east, 20′ south of A and 10′ below A. Point C is 25′ west, 25′ south of A and 25′ below A. Scale: $1″ = 20′$. Using the revolution method, determine the angle between the two sewer lines.   *Ans.* Angle $= 34°30′$

28. The axis of a shaft is represented by line AB in Fig. 7-26. Point X is the tip of a lever handle perpendicular to the axis. Point B is 8′ east, 3′ south of A and 6′ above A. Point X is 3′ due east of A and at the same elevation as B. Scale: $\frac{1}{4}″ = 1′–0″$. If point A is on the floor, how much clearance, if any, is there between the lever handle and the floor? How long is the lever handle? If there is a clearance, how long can the handle be made in order to just touch the floor?

Ans. Clearance $= 6″$
Handle Length $= 3′–4\frac{1}{2}″$
Maximum Length $= 4′–1\frac{1}{2}″$

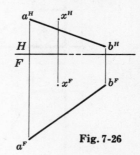

Fig. 7-26

29. Fig. 7-27 shows the structural framework between two adjacent buildings. Scale: $\frac{1}{4}″ = 1′–0″$. By means of revolution, determine the true length of structural members AB and CD.
Ans. T.L. $AB = 11′–3″$, T.L. $CD = 13′–3″$

30. Two pipelines AB and CB converge at point B. Point A is 35′ north, 10′ east of B and 30′ above B. Point C is located 20′ north, 60′ west of B and 15′ above B. A new pipeline DB is to be located in the plane of ABC connecting B and point D which is located 30′ due west of A. Scale: $1″ = 20′$. By means of revolution, determine the true length of each pipeline.
Ans. T.L. $AB = 47′–0″$, T.L. $CB = 64′–6″$, T.L. $DB = 49′–0″$

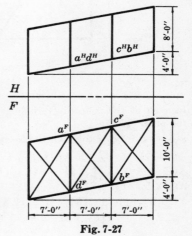

Fig. 7-27

31. Plane ABC and point X. Point B is 2′ east, 2′ north of A and 2′ below A. Point C is 3′–6″ east, 9″ north of A and 9″ above A. Point X is 2′ east of A and 1′ north of A. Scale: $1″ = 1′–0″$. Using the revolution method, show a 9″ diameter hole on the plane and having X as its center. How close does the circle come to the nearest edge of the limited plane?   *Ans.* $5\frac{3}{4}″$

32. Planes ABC and BCD are shown in Fig. 7-28. Point B is 11′ east, 6″ north of A and 1′–6″ below A. Point C is 3′ east, 5′–6″ south of A and 3′ above A. Point D is 6′–6″ east, 3′–6″ north of A and 8′ above A. Scale: $\frac{1}{4}″ = 1′–0″$. Determine the true size of the dihedral angle by revolution.   *Ans.* $78°30′$

33. ABC and ABD are two planes which intersect along line AB. Point B is 4′ east, 8′ south of A and 8′ below A. Point C is 11′ east, 2′ north of A and 10′ below A. Point D is located 7′ due north of A and 11′ below A. Scale: $\frac{1}{8}″ = 1′–0″$. By means of revolution, determine the dihedral angle between the two planes.
Ans. Dihedral Angle $= 50°$

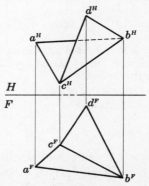

Fig. 7-28

34. Fig. 7-29 shows the plan and front elevation views of a concrete bridge pier. Scale: $\frac{3}{16}'' =$ $1'-0''$. By means of revolution, determine the true size of the dihedral angle between (a) planes B and C, (b) planes A and B.

   Ans. (a) Angle $= 95°$
        (b) Angle $= 100°$

35. Fig. 7-30 shows two intersecting planes, A and B. Scale: $\frac{1}{4}'' = 1'-0''$. By means of revolution, determine the dihedral angle between the two planes.   Ans. Angle $= 99°$

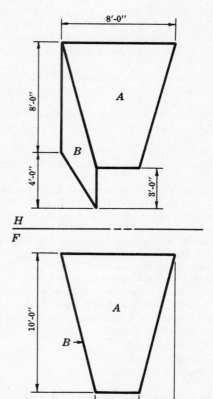

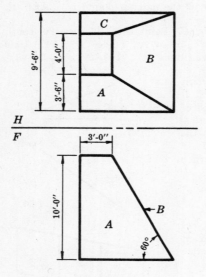

Fig. 7-29

Fig. 7-30

   In each of the following problems determine, by revolution, the true size of the angle between line XY and the plane ABC.

   See Art. 1.7 for the coordinate system of problem layout.

36. $X(1, 2, 5)$ $Y(3, 2, 5\frac{1}{4})$ $A(1\frac{1}{2}, 1\frac{1}{2}, 6\frac{1}{2})$ $B(2, 3, 5\frac{1}{4})$ $C(4, 2, 5)$ Ans. $26°30'$
37. $X(1\frac{1}{2}, \frac{1}{2}, 4\frac{1}{2})$ $Y(3, 2, 4\frac{1}{2})$ $A(1, 1\frac{1}{2}, 5)$ $B(2, 2\frac{1}{2}, 4)$ $C(3\frac{1}{2}, \frac{1}{2}, 5\frac{1}{2})$ Ans. $38°30'$
38. $X(1, 2, 5)$ $Y(3\frac{1}{2}, 2\frac{1}{4}, 4)$ $A(1\frac{1}{2}, 1\frac{1}{2}, 5)$ $B(2, 3\frac{1}{4}, 4\frac{1}{2})$ $C(3\frac{1}{2}, 1, 6)$ Ans. $37°30'$
39. $X(1\frac{1}{2}, 3\frac{1}{4}, 5\frac{1}{4})$ $Y(3, 3\frac{1}{4}, 6\frac{1}{2})$ $A(1, 3\frac{1}{2}, 6\frac{1}{2})$ $B(2\frac{1}{4}, 2\frac{1}{2}, 5\frac{1}{4})$ $C(3\frac{1}{2}, 4\frac{1}{4}, 6\frac{1}{4})$ Ans. $43°$
40. $X(2, 2\frac{1}{4}, 4\frac{1}{2})$ $Y(1\frac{1}{2}, 3, 6)$ $A(\frac{1}{4}, 3\frac{1}{2}, 5\frac{1}{2})$ $B(1\frac{1}{2}, 3\frac{1}{2}, 7)$ $C(1\frac{1}{4}, 1\frac{1}{2}, 4\frac{1}{2})$ Ans. $22°30'$

41. Plane ABC and the line XY are located as follows: $A(5\frac{1}{4}, 2\frac{1}{4}, 6\frac{1}{4})$ $B(6, \frac{1}{2}, 5\frac{1}{4})$ $C(6\frac{1}{4}, 1\frac{1}{2}, 6\frac{1}{4})$ $X(5\frac{1}{4}, 1\frac{1}{2}, 7)$ $Y(6, 2\frac{1}{4}, 5\frac{1}{2})$. By means of revolution, determine the true size of the angle between line XY and the plane ABC.   Ans. Angle $= 76°$

42. Line AB is located as follows: Point B is 2' east, 1' south of A and $1'-6''$ above A. Scale: $1'' = 1'-0''$. Determine, by means of revolution, the true length of the line and the angle it makes with the H, F, and P planes.

   Ans. T.L. $AB = 2'-8\frac{3}{8}''$
         Angle $H = 34°$
         Angle $F = 22°$
         Angle $P = 48°$

43. A pipe XY and a wall $ABCD$ are located as follows: $X(1\frac{1}{2}, 5, 7\frac{1}{2})$ $Y(3, 5\frac{1}{2}, 7\frac{1}{2})$ $A(2, 6, 7)$ $B(3, 4\frac{1}{2}, 8)$ $C(4\frac{1}{2}, 4\frac{1}{2}, 7\frac{1}{2})$ $D(3\frac{1}{2}, 6, 6\frac{1}{2})$. By means of revolution, determine the angle between the pipe and the plane of the wall.   Ans. Angle $= 26°$

44. Fig. 7-31 below shows a tetrahedron. Scale: $12'' = 1'-0''$. By means of revolution, determine the angle the line *BC* makes with the plane *ABD*.   *Ans.* Angle = 51°

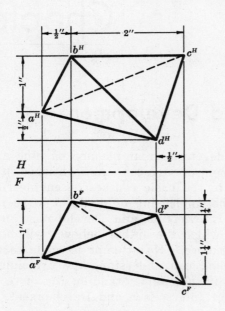

Fig. 7-31

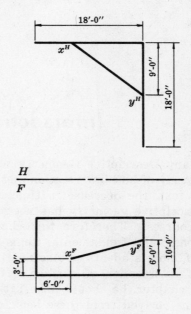

Fig. 7-32

45. Two vertical walls are intersected by a pipe, *XY*. See Fig. 7-32 above. Scale: $\frac{1}{8}'' = 1'-0''$. By means of revolution, determine the angle that the pipe makes with the wall (*a*) at point *X*, (*b*) at point *Y*.
Ans. (*a*) Angle = 36°, (*b*) Angle = 51°30′

Chapter 8

Intersection and Development

Many Descriptive Geometry textbooks devote a great amount of time and material to the study of both intersection and development of surfaces. As mentioned in the preface to this text, the decrease in time allocated to the basic courses of engineering drawing and descriptive geometry has necessitated the condensing of material to be presented in these courses. Therefore this chapter on Intersection and Development presents only those basic and fundamental items of study which the author believes will best meet the need of most students of engineering and science. No attempt is made to reach every type of intersection of surface. If the student can understand the principles explained in this chapter, he should have little or no difficulty understanding the more complicated problems derived from these basic concepts of Intersection and Development.

In previous chapters we have explained the various methods involved in the determination of intersecting lines, planes, and combinations of lines and planes. However, the study of intersecting lines and planes reaches a more practical realm when the student can see the application of these aforementioned methods. The purpose of this chapter is to analyze the more common types of surface intersections which are more likely to be of direct benefit to the practicing engineer. The examples are such that the fundamental principles of surface intersections can be readily adhered to without complicated descriptive steps of procedure. The problems in this chapter are basic and yet involve careful analysis on the part of the individual student.

In Chapter 4 we have shown that a line intersects a plane at a point which is common to both the line and the plane. We have shown also that the intersection of two nonparallel planes is a straight line common to both planes. The author suggests that the student return to Articles 4.1 and 4.2 to refresh his memory regarding the basic concepts of intersecting lines and planes.

8.1 DEFINITIONS

(1) *Generatrix* — a straight line whose continuous motion generates, or forms, a surface.

(2) *Directrix* — a straight or curved line continuously in contact with the generatrix.

(3) *Director* — a surface to which the generatrix is constantly parallel.

(4) *Element* — a straight line shown on the surface indicating a specific position of the generatrix.

(5) *Axis* — a centerline about which a generatrix revolves.

(6) *Development* — laying out, or unfolding, a surface into a plane. Fig. 8-1(*a*) and (*b*) below show the theory behind the development of a cylinder and cone respectively. A developed surface shows all lines in their true length and all angles in their true size. The surface is usually assumed to be cut at its shortest element to facilitate ease in fabrication. In the problems to follow, the additional surface necessary for "lapping" or "seaming" is not shown.

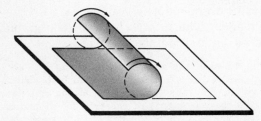

Fig. 8-1(*a*). Development of a Right Cylinder   **Fig. 8-1(*b*). Development of a Right Circular Cone**

(7) *Triangulation* — a method used in the development or approximate development of surfaces whereby the surfaces are divided into triangles. Once the true length of each side of each triangle is obtained, all the triangles are then connected in proper sequence to form the development.

(8) *Ruled Surface* — a surface generated by a straight line.

(9) *Single-Curved Surface* — a ruled surface which can be developed (e.g. cone, convolute, cylinder).

(10) *Double-Curved Surface* — a surface generated by a curved generatrix which revolves about an axis or moves along a curve. It has no straight line elements (e.g. sphere, torus, paraboloid).

(11) *Warped Surface* — a ruled surface which cannot be developed. No two consecutive elements may be parallel or intersect (e.g. conoid, cylindroid, helicoid).

(12) *Right Section* — a plane section perpendicular to the axis.

(13) *Girth Line or Stretchout Line* — a line drawn perpendicular to the true length elements in a view showing the development.

(14) *Bend Line* — lateral edge along which the development is folded to form the desired shape.

INTERSECTION

8.2 To DETERMINE the POINTS at WHICH a STRAIGHT LINE PIERCES a CYLINDER

Analysis: Right Circular Cylinder. Draw a view showing the axis of the cylinder as a point. Show the straight line in this same view and the points of intersection will be obvious. These points can be located in other views by simple projection.

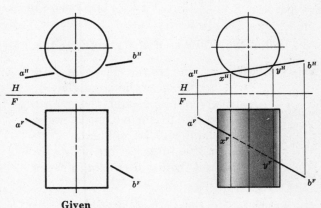

Given

Fig. 8-2. Line Intersecting a Right Circular Cylinder

Example: In Fig. 8-2 the plan and front elevation views of the line and cylinder are given. Points X and Y in the plan view obviously determine the location of the intersection. Project X and Y down to the front elevation views and determine proper visibility.

Analysis: Oblique Cylinder. If successive views are drawn to show the axis of the cylinder as a point, then the problem is reduced to the analysis above. However the two-view cutting-plane method is usually employed in the case of an oblique cylinder because it is simple enough to be readily understood and is quicker.

A cutting plane containing the given line and parallel to the axis of the cylinder will intersect the cylinder in elements. The intersections of the given line with these elements will be the required "pierce points".

Example: In Fig. 8-3 below, the plan and front elevation views of the oblique cylinder and line are given. A cutting plane containing line AB and parallel to the axis of the cylinder is drawn in both views. This is done by drawing line BC parallel to the axis in both views. The upper base plane of the cylinder is intersected by the cutting plane in the line CD, cutting across the upper base of the cylinder at points V and Z. These points V and Z lie in the cutting plane and are the higher ends of the straight-line elements. One straight-line element is drawn from V in both views and the other element from Z is drawn in the plan view. The front view of the element from Z happens to be the extreme element shown in the front. Both elements must be drawn parallel to the axis of the cylinder. The intersection of these elements with the line AB determines the pierce points X and Y. These pierce points can be verified by projecting between views.

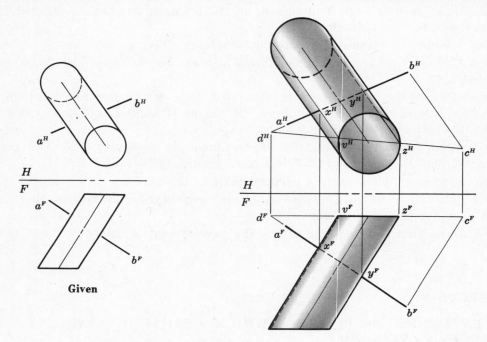

Fig. 8-3. Line Intersecting an Oblique Cylinder

8.3 INTERSECTION of a PLANE and a PRISM

Analysis: Draw an auxiliary elevation view which will show the edge view of the given plane, and the points where the prism edges intersect the plane will be evident. Connect these intersection points in plan and front views using proper visibility.

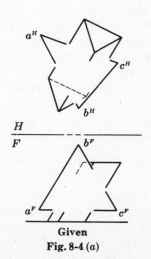

Given
Fig. 8-4 (a)

Example: Fig. 8-4(a) shows the plan and front elevation views of plane ABC and the oblique prism. Draw auxiliary elevation view 1 to show the edge view of plane ABC and the prism [see Fig. 8-4(b)]. The parallel edges of the prism will intersect the plane at points 1, 2, and 3. Project these three points back to the plan and front views and connect them, showing proper visibility for plane and prism.

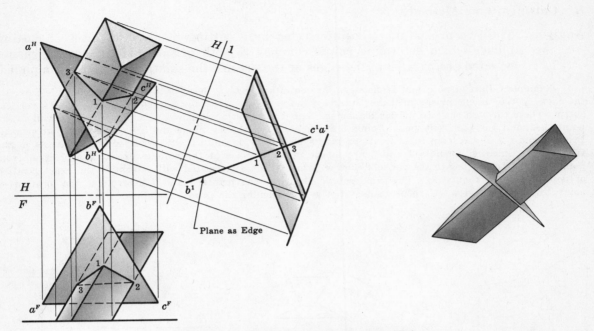

Fig. 8-4 (b). Intersection of a Plane and a Prism

8.4 INTERSECTION of PRISMS

A. Edge View Method

Analysis: Determine where the edges of the two prisms pierce each other by showing the edge view of each prism. Connect the piercing points and determine the proper visibility.

> *Note:* If only one prism appears as an edge in one of the given views, a new auxiliary view can be drawn to show the edge view of the other prism.

Example: In Fig. 8-5, the extreme surface limits of the two prisms are given in the three primary views. Extend the surface limits of the horizontal prism in the plan view until they intersect the edge view of the vertical prism. Extend the surface limits of the vertical prism in the profile view until they intersect the edge view of the horizontal prism. Label the piercing points as shown in the figure. Project the piercing points from the plan view down to the front view until they meet the projection of corresponding points from the profile view. Careful visualization will determine proper visibility.

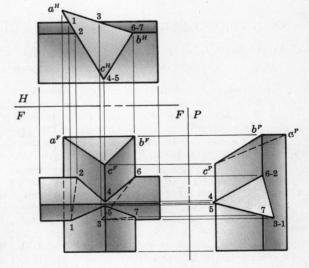

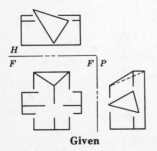

Given

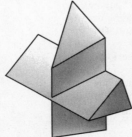

Fig. 8-5. Intersection of Prisms (Edge View Given)

B. Cutting-Plane Method

Analysis: The two prisms must be intersected by a cutting plane or a series of cutting planes which determine points common to both prisms. If the cutting planes selected are parallel to the edges of the prisms, the solution is further simplified.

Example: Partial plan and front elevation views are given. See Fig. 8-6. In the plan view extend the edges of the oblique prism until they intersect the vertical surfaces of the right prism. Pass a vertical cutting plane through the two prisms as shown. Label the points as shown. Simple projection will locate points 1, 2, and 3 in the front view. Points 6 and 7 lie in the cutting plane and are common to both prisms. Project points 4 and 5 from their locations in the plan view down to their corresponding positions in the front view. Since both points 4 and 6 are in the upper surface of the oblique prism, a projection from point 4 in the front view to the vertical edge on which point 6 is located will determine the exact position of point 6 in the front view. The same relationship between points 5 and 7 determines the location of point 7 in the front view. Careful visualization will determine the proper visibility.

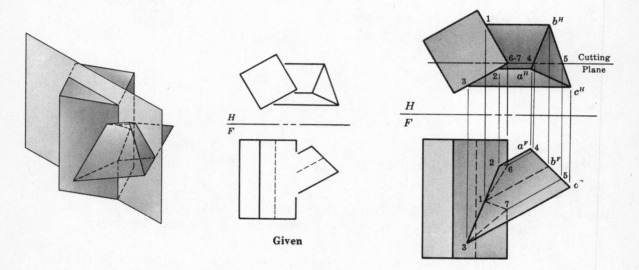

Fig. 8-6. Intersection of Prisms (Cutting-Plane Method)

8.5 INTERSECTION of PRISM and PYRAMID

Analysis: Draw an auxiliary view which will show the prism as an edge. Use the pierce point method to locate the points common to both prism and pyramid.

Example: Partial plan and front elevation views given in Fig. 8-7(*a*). Draw auxiliary elevation view to show the pyramid and the edge view of the prism. Refer to Fig. 8-7(*b*). Label points 1 through 6 on the edge view of the prism in view 1. Points 3 and 4 can be projected directly to the plan view and subsequently to the front elevation view. Pass an edgewise cutting plane through points 1-5 and 2-6 in view 1. This cutting plane will intersect *OC* at point 7, *BC* at point 8, and *OA* at point 9. Project lines 7-8 and 7-9 to the plan view to locate points 1, 2, 5, and 6 in the plan view. Simple projection will locate these points in their corresponding positions in the front view. A further check for the plan view of point 2 is made by passing a cutting plane through points 3 and 2 in view 1 until it intersects *OB* at point 10. Project point 10 to the plan view and draw a line from point 3 to point 10. If drawn correctly, the line will pass through point 2 in the plan view. Careful visualization will determine proper visibility.

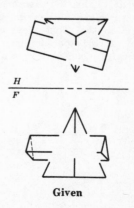

Given

Fig. 8-7 (*a*)

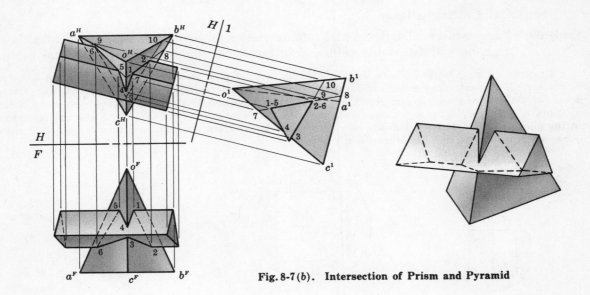

Fig. 8-7(b). Intersection of Prism and Pyramid

8.6 INTERSECTION of CONE and PRISM (CUTTING-PLANE METHOD)

A. Vertical Cutting Planes

Analysis: A series of vertical cutting planes passing through the axis of the cone and cutting the prism will have elements on which will be a point of intersection common to both cone and prism.

Example: Fig. 8-8 below shows the given views. In the plan view, pass a series of cutting planes through the axis of the cone which intersect the edge view of the given prism. Label these points 0 through 6 and *A* through *E* as shown. Show the elements in the front view. Project points *A*, *B*, *D*, and *E* to the front view until they intersect elements 1, 2, 4, and 5 respectively. Point *C* in the front view will be at the same elevation as the highest point of intersection of the cone and prism shown by the extreme elements of the cone in the front view. Connect the points *A* through *E* to show proper visibility for the line of intersection.

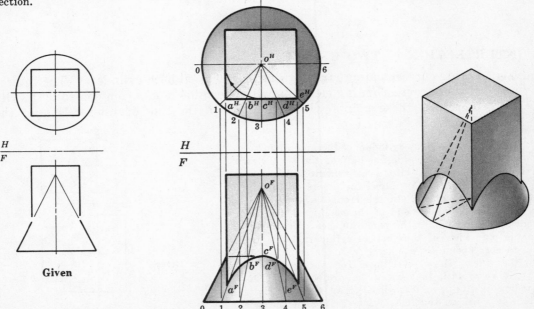

Fig. 8-8. Intersection of Cone and Prism (Vertical Cutting Planes)

B. Horizontal Cutting Planes

Analysis: A series of parallel horizontal cutting planes perpendicular to the vertical axis of the cone will determine points of intersection common to both cone and prism.

Example: Fig. 8-9 below shows the given views. In the plan view draw circles 1, 2, and 3, being sure to include the extreme elements of the edge view of the prism. Draw the horizontal cutting planes in the front elevation view. In the plan view, label the intersection of the cutting planes with the prism as points A through E. Project these points down to the front view until they meet the corresponding cutting planes. Points A through E are now determined in the front view and are then connected to show proper visibility of the line of intersection.

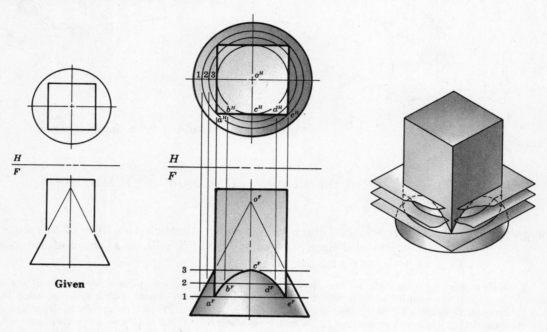

Fig. 8-9. Intersection of Cone and Prism (Horizontal Cutting Planes)

8.7 INTERSECTION of TWO CYLINDERS

Analysis: Locate corresponding straight line elements on both cylinders. The intersection of these corresponding elements will determine points common to both cylinders; therefore these points will be located on the line of intersection of the two cylinders.

Example: Two right cylinders with same diameter given in Fig. 8-10(a). Divide one-half of the edge view of the vertical cylinder into a convenient number of segments. See Fig. 8-10(b). Label the points on the circumference 1 through 13. In the front elevation view, draw a semi-circle at the end of the horizontal cylinder as shown. Divide it into the same number of segments as in the plan view of the vertical cylinder. Label the points to correspond elevationally with the same points in the plan view. Draw level lines from these points in the front view until they meet the projection of corresponding points from the plan view. Connect these points which are on the line of intersection of the two cylinders to obtain proper visibility. Symmetry determines the complete line of intersection.

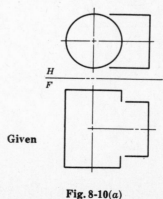

Given

Fig. 8-10(a)

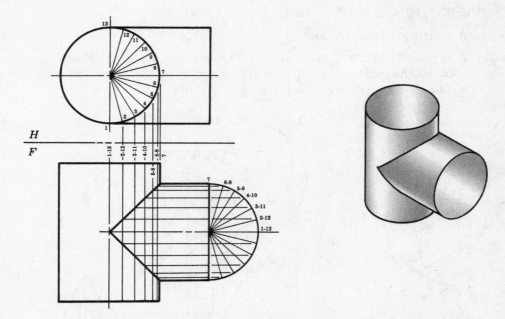

Fig. 8-10(*b*). Intersection of Cylinders — Same Diameters

Example: Two right cylinders with different diameters. Refer to Fig. 8-11 below. Draw a revolved right section of the inclined cylinder in both the plan and front elevation views. Divide the right sections into a convenient number of elements. Label each element, making sure that their corresponding positions are orthographically correct. In this case, element 5 is high and element 13 is low. Element 1 is farthest away from the observer and element 9 is closest to the observer. Show all elements of the inclined cylinder in both views, locating them parallel to the axis of the inclined cylinder. In the plan view, where the elements intersect the edge view of the vertical cylinder, label these points with their proper element number. Project each intersection point on the plan view down to the front view until it meets its corresponding element in this view. These "meeting points", such as 8, 9, and 10, are points on the line of intersection of the two cylinders. Connect these points and be careful to obtain proper visualization.

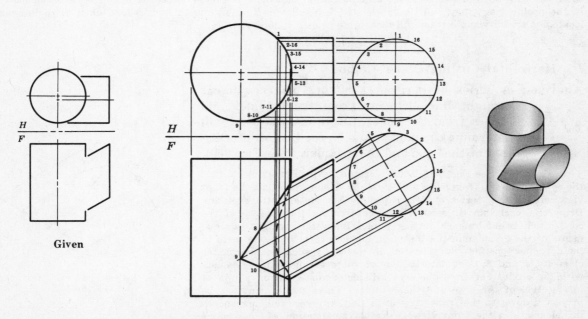

Fig. 8-11. Intersection of Cylinders — Different Diameters

8.8   INTERSECTION of a CYLINDER and a CONE

A.   Inclined Cutting-Plane Method

Analysis: A series of cutting planes passing through the vertex of the cone and the edge view of the cylinder will locate the common points of intersection between the cone and the cylinder.   See Fig. 8-12 below.

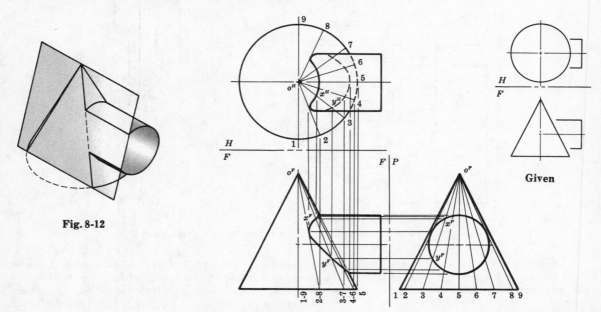

Fig. 8-12

Fig. 8-13.   Intersection of Cylinder and Cone (Inclined Cutting Planes)

Given

Example: See Fig. 8-13 above.   Draw a profile view to show the cone and the edge view of the cylinder.   Pass a series of cutting planes through the vertex O and cutting the edge view of the cylinder. Label the elements on the baseline of the cone 1 through 9.   Show these elements in both plan and front elevation views.   From each point of intersection shown on the profile view, such as X and Y, project to the front view until it intersects the corresponding element in this view.   Because X and Y lie on element 3 in the profile view, they will lie on element 3 in all views.   Continue this procedure until all points are located.   Careful visualization will determine proper visibility.

B.   Horizontal Cutting-Plane Method

Analysis: A series of horizontal cutting planes will cut straight-line elements from the cylinder and circles from the cone.   Their intersections will determine common points on the line of intersection of the cone and the cylinder.   See Fig. 8-14.

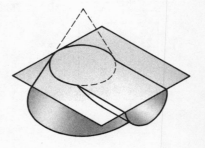

Example: See Fig. 8-15 below.   Draw a revolved right section of the cylinder in the front view and divide it as shown so that the front view will show a series of seven parallel horizontal cutting planes. Draw the circles in the plan view, representing the surface of the cone which is cut by each cutting plane.   Show the straight-line elements of the cylinder in the top view, being careful to locate them in their proper position by transfer of distances such as shown for elements 2 and 6.   The intersection of these straight-line elements with the circles cut from the cone will determine common points of intersection of the cone and the cylinder.   These points can then be projected down to the front view until they intersect the corresponding elements in the front view.   Careful visualization will determine proper visibility.   Symmetry determines the complete line of intersection.

Fig. 8-14

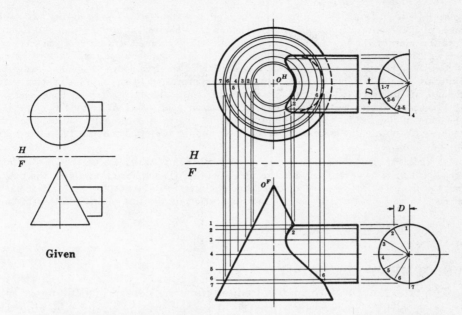

Fig. 8-15. Intersection of Cylinder and Cone (Horizontal Cutting Planes)

8.9 INTERSECTION of TWO RIGHT CIRCULAR CONES HAVING PARALLEL BASES

Analysis: A series of cutting planes parallel to the bases of the cones will intersect each of the cones in a circle. The intersection of the circles will determine points which are common to the line of intersection of the two given cones.

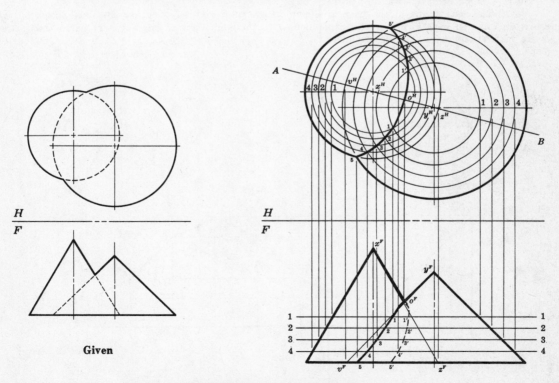

Fig. 8-16. Intersection of Two Right Circular Cones

Example: In Fig. 8-16 above, the partial plan and front elevation views of the two cones are given. Since the bases of both cones are horizontal, a series of four horizontal cutting planes are passed through the front elevation view of the intersecting cones. The cutting planes cut a circle from each cone as shown in the plan view. The intersection of the corresponding circles from each cone will determine points common to each cone and these points lie on the line of intersection of the two cones. Continue the process until all required points are established in the plan view and then projected to the front elevation view. To locate the highest point of intersection, point O, draw cutting plane AB through both cone vertices, X and Y. This cutting plane determines two straight-line elements in each cone, XZ and YV. The intersection of these two elements in the front view will determine point O, the highest point on the line of intersection of the two cones. To obtain the plan view of point O simply project from the front view to the cutting plane AB in the plan view.

As far as visibility is concerned, in the plan view all points which lie on the line of intersection are visible. In the front view however, points 1 through 5 will be visible since they lie on the near half of each corresponding set of circles cut from each cone. Since points 1' to 5' lie on the far half of each corresponding set of circles, these points will be on the hidden portion of the line of intersection in the front view.

DEVELOPMENT

8.10   DEVELOPMENT of a PRISM

Analysis: The true length of the lateral edges must be determined along with their relative positions in respect to the right section of the prism. A view showing the right section of the prism will also determine the length of the development.

Example: The plan and front elevation views of the truncated hexagonal prism are given. See Fig. 8-17 below. Label the vertical edges of the prism as shown. Draw a girth line (stretchout line) directly across from the hexagonal base of the prism. The length of the girth line must be equal to the perimeter of the prism base. Along this girth line, lay out the distances AB, BC, etc., taken from the plan view where these distances appear in their true length. Draw the lateral edges perpendicular to the girth line and equal to their true length shown in the front view. Connect points 1, 2, 3, etc. with straight lines to complete the development of the lateral surfaces of the prism. The "bend lines" are shown by a circle over the line to be bent to establish proper shape.

Note: The true shape of the base is shown in the plan view. To obtain the true shape of the cut surface, an inclined view is projected perpendicular to the edge view of the cut surface in the front view and can be transferred to the development.

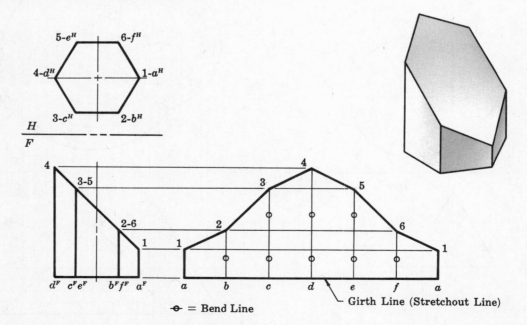

⊸ = Bend Line

Girth Line (Stretchout Line)

Fig. 8-17. Development of a Prism

8.11 DEVELOPMENT of a RIGHT PYRAMID

Analysis: Determine the true length of each of the lateral edges which radiate from a vertex point. Knowing these true lengths and the perimeter of the pyramid base is sufficient to complete the development.

Example: See Fig. 8-18 below. The plan and front elevation views of the pyramid are given. Label the lateral edges as shown. All inclined edges of the right pyramid are of equal length, and it is, therefore, necessary to find the true length of only one edge. The true length of the base perimeter is seen in the plan view. Revolve edge *OD* in the plan view to obtain the true length in the front view. This same distance "X" will also be the true length of *OA*, *OB*, and *OC*. Using distance *X* as a radius, swing an arc of indefinite length. From a point *D* on the arc, step off a chordal distance equal to distance *DA* in the plan view. This locates point *A* in the development. Continue this procedure until the development is complete.

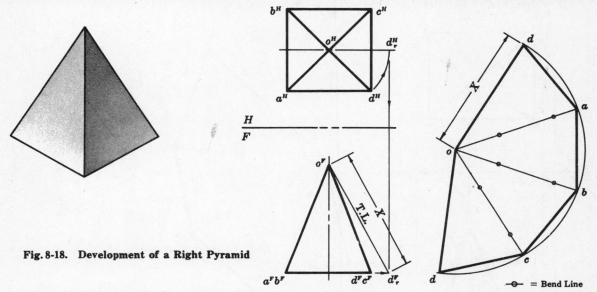

Fig. 8-18. Development of a Right Pyramid

Example: (Truncated Right Pyramid.) See Fig. 8-19 below. The plan and front elevation views are given. Label the edges as shown. Since the cut surface appears as an edge in the front view, edges *A*-1 and *B*-2 are equal and *C*-3 is the same length as *D*-4. It is necessary, therefore, to determine the true length of two edges — in this case, *A*-1 and *D*-4. By revolution, these two true length distances are measured in the front view. Using the true length distance *OD* as a radius, swing an arc of indefinite length. Along line *OD*, measure the true length of *D*-4 as obtained in the front view. From point *D*, swing an arc equal to *DA* in the plan view and intersecting the arc of indefinite length. This locates point *A* on the development. Draw a line from *A* to point *O*. Along this line, measure the true length of *A*-1 as obtained in the front view. Continue this procedure until the development is complete. The true size of the base is shown in the plan view. An inclined view 1 will determine the true size of the cut surface.

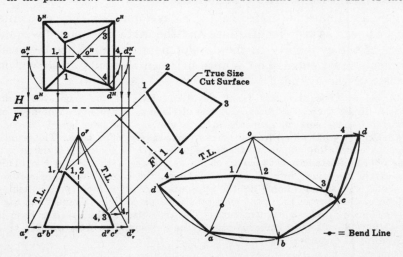

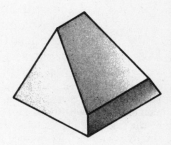

Fig. 8-19

**Development of a
Truncated Right Pyramid**

8.12 DEVELOPMENT of an OBLIQUE PYRAMID

Analysis: The procedure used to develop an oblique pyramid is essentially the same as for a right pyramid except that the lateral edges of an oblique pyramid are of various lengths; therefore, the true length of each edge must be individually determined. Usually the true length of each edge can best be obtained by the method of revolution.

Example: See Fig. 8-20 below. The plan and front elevation views of the oblique pyramid are given. Label the edges as shown. Revolve each edge to obtain its true length in the front view. For the development, lay out *OA* equal to the true length of *OA* in the front view. From point *A*, swing an arc equal to *AB* in the plan view. From point *O*, swing an arc equal to the true length of *OB* in the front view until it intersects the arc from *A* drawn previously. This intersection point will determine point *B* in the development. Continue the procedure until the development is complete.

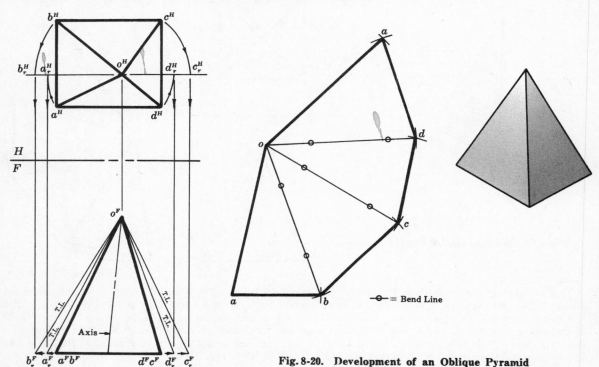

Fig. 8-20. Development of an Oblique Pyramid

8.13 DEVELOPMENT of a RIGHT CIRCULAR CYLINDER

Analysis: Fig. 8-1(*a*) shows how a cylinder is "rolled out" for development. The elements of the cylinder must appear as true length lines and the girth line must be equal to the circumference of the cylinder. In the development, the distance between the true length elements must equal the distance between elements as shown on the right section view of the cylinder.

Example: Refer to Fig. 8-21 below. The plan and front elevation views of the truncated right cylinder are given. In the plan view, divide the circumference of the circle into a convenient number of parts — in this case, 16. Show and label these vertical elements in both views. The true length of each element is shown in the front view. Draw a girth line directly across from the cylinder base. The length of the girth line is established either by calculating the circumference equal to πD, or by using dividers and simply stepping off the sum of the chordal distances between elements as shown in the plan view. This latter method is preferred. If the circle is divided into 16 parts as shown, the chordal distance method results in the negligible error of about 0.6 percent. Label all the elements and draw them perpendicular to the girth line, making them as long as they are in the front view by simple projection. Connect the high points of each element in the development by using a French curve. The true size of the cylinder base is shown in the plan view. An inclined view 1 will show the true size of the cut surface. Attach these true size surfaces to the surface development if required.

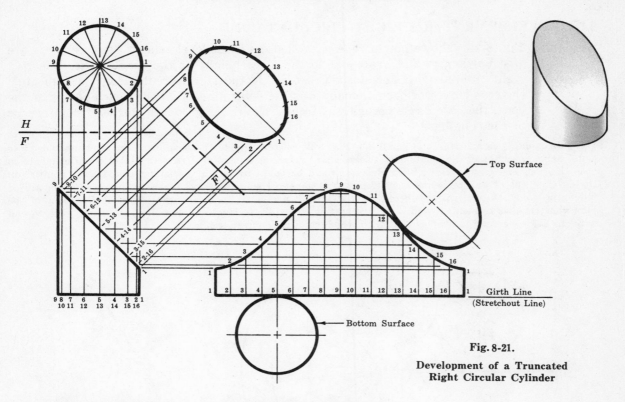

Fig. 8-21.

Development of a Truncated Right Circular Cylinder

8.14 DEVELOPMENT of an OBLIQUE CYLINDER

Analysis: A view should be drawn to show the true length of the cylinder elements. Another view is then projected showing the right section of the cylinder and the true distance between the elements which will appear as points in this view.

Example: See Fig. 8-22 below. The plan and front elevation views of the oblique cylinder are given. Draw an auxiliary elevation view of the cylinder showing the axis XY in its true length. Show a revolved right section of the cylinder and divide half of the elliptical circumference into six equal divisions as shown. Label these elements 1 through 7 and project them back to the elevation view 1 where they will appear in their true length and parallel to the axis XY. Draw the girth line perpendicular to the true length of the axis and directly across from element 1. Along the girth line step off six spaces, each equal to the chordal distance between points 1-2 or 2-3, etc. as shown in the revolved right section of the cylinder. Draw the element lines perpendicular to the girth line. From view 1 project the extreme points of the elements 1 through 7 until they intersect their corresponding element lines in the development. Connect these points of intersection by using a French curve. Since the cylinder is symmetrical, a half development is sufficient to explain the method involved.

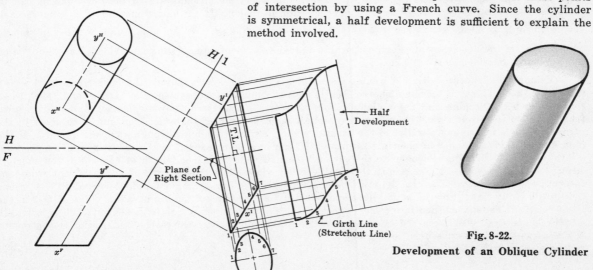

Fig. 8-22.

Development of an Oblique Cylinder

8.15 DEVELOPMENT of a RIGHT CIRCULAR CONE

Analysis: Fig. 8-1(*b*) shows how a cone is "rolled out" for development. All elements are of equal length and are equal to the slant height of the cone if the cone is a right circular cone and not truncated. If the right circular cone is truncated, the true length of each element must be determined. In the development of the cone, the base circle becomes a circular arc equal in length to the circumference of the cone base.

Example: (Right Circular Cone.) See Fig. 8-23 below. Divide the base circle in the plan view into equal parts to establish the location of twelve equally-spaced elements. Draw and label these elements in both the plan and front views. Each element would be equal in length to distance *A*. Locate point *O* in the development at any convenient position on the paper. Using distance *A* as a radius, swing an arc of indefinite length. Begin with element 1 and swing an arc equal to the chordal distance *B* shown in the plan view between elements 1 and 2. Continue this operation to locate all the elements, being sure to end with the same element number as that with which the development was started.

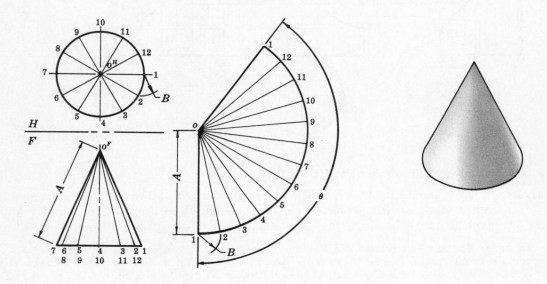

Fig. 8-23. Development of a Right Circular Cone

Note: The graphical method explained is usually satisfactory, and dividing the circular base into more parts will yield a greater degree of accuracy. However, the circular arc in the development can be computed mathematically if it is necessary for greater precision. If the latter method is chosen, the following formula is used: $\theta = \dfrac{R}{S}(360°)$ where *R* equals the radius of the base and *S* equals the slant height of the cone. If the development is symmetrical, a half development can be shown provided that the division is made through a symmetrical axis.

Example: (Truncated Right Circular Cone.) See Fig. 8-24 below. Divide the base circle in the plan view into 12 equal parts to establish the location of 12 equally-spaced elements. Draw and label these elements in both the plan and front views. Since the cone is truncated, only elements 1 and 7 are shown true length in the front view. Therefore, the highest point on each of the other elements is revolved in the plan view to lie in a plane parallel to the frontal plane. Assume a plane through elements 1 and 7 in the plan view. The true length of elements 2 through 12 can now be measured in the front view after drawing a level line from their highest point to the extreme element *O*-1. The true length of element 1, the shortest element, would be equal to *O*-1 minus *O*-*B* in the front view. The true length of each element would be determined in this way. The true length of elements 4 and 10, for example, would be equal to *O*-1 minus *O*-*C* in the front view. For the actual development, draw a circular arc having a radius equal to *O*-1 in the front view and equally divided into twelve spaces, each having a distance of *A* as shown. The true length of each element derived from the front view will be laid out on each corresponding radial line element. A French curve will be useful in laying out a smooth development.

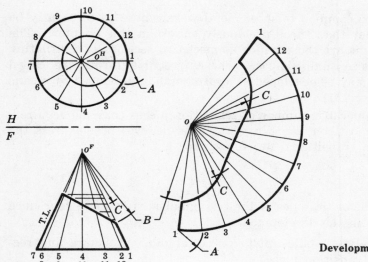

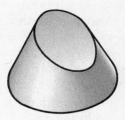

Fig. 8-24.
Development of a Truncated Right Circular Cone

8.16　DEVELOPMENT of an OBLIQUE CONE

Analysis: The elements will be unequal in length, and therefore the true length of each element must be determined. The cone base will determine the length of the development, but the base will not develop as a circular arc such as seen in Article 8.15.

Example: Fig. 8-25. The plan and front elevation views of the oblique cone are given. In the plan view, divide the circumference of the base into twelve parts. Show these twelve elements in both views. In the plan view revolve the elements into a frontal plane in order to obtain their true length as shown in the true length diagram. Draw a line O-1 to start the development. From point 1, swing an arc equal to the chordal distance "A" in the plan view between the elements at the base of the cone. From point O, swing an arc equal to the true length of O-2 as measured in the true length diagram. The intersection of this arc with the arc from point 1 will determine point 2 in the development. Continue this procedure until the development is complete. Connect the points with a smooth curve.

Note: If the oblique cone had been truncated, the procedure would be the same except that the highest point on each element would be projected from the front view to the true length diagram. In this case, the cone should be cut along the shortest element line.

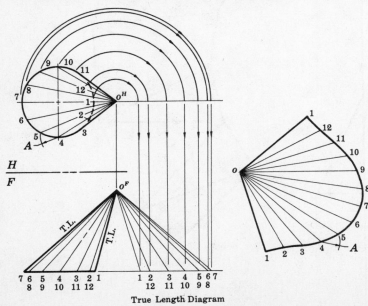

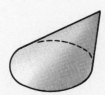

True Length Diagram

Fig. 8-25.
Development of an Oblique Cone

8.17 TRANSITIONS

There are many irregular forms common to sheet metal work whose patterns may be developed by means of methods other than those previously shown in this chapter. The most common types of irregular forms are those which connect two sections of ventilating or heating ductwork of different sizes and shapes. These irregular forms are called transition pieces, and their surfaces may be developed or approximately developed by means of triangulation (see definitions, Page 131).

Since it is obvious that an innumerable number of transition pieces could be required, the author has selected two of the most common types which he believes best express the basic principles involved in all such transitional developments.

A. Transition — Square to Square

Analysis: Determine the true lengths of the sides of the triangular surfaces and reproduce them in proper sequence in the development.

> *Note:* The same analysis would apply for rectangular to square and rectangular to rectangular transition pieces.

Example: In Fig. 8-26 below, the plan and front elevation views of a transition piece are given. The true size of the square base and top are shown in the plan view. At the side of the front elevation view construct a true length diagram to obtain the true length of the seam line 3-E and also the true length of the bend line 3-C. Since all of the lines connecting the corners in the plan view are equal, the true length distance obtained for 3-C would also apply for 2-B, 2-C, 1-A, 1-B, etc.

To lay out the complete development in one piece, lay out the true length of the seam line 3-E. From point 3 swing an arc equal to the true length of 3-C. From point E swing an arc equal to the true length of E-C which can be obtained from the plan view. The intersection of these two arcs determines the location of point C in the development. Similarly, the intersection of the arcs 3-2 drawn from point 3, and C-2 drawn from point C determines point 2. Continue this procedure until the seam line is again drawn at the other end of the development.

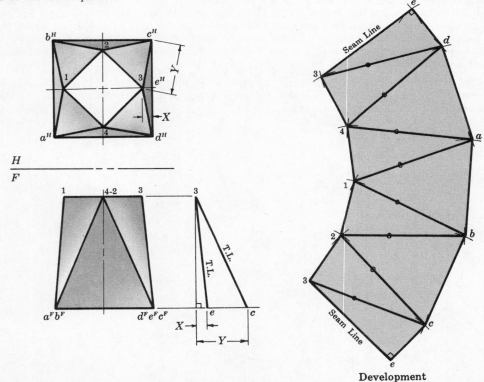

Development

Fig. 8-26

B. **Transition — Rectangular to Circular**

Analysis: Divide the surface of the transition into triangular areas and determine the true length of each side of the triangle as in the previous analysis. Actually it involves determining the true length of the bend lines required to form the metal from the rectangular shape to the circular shape. Once the triangles are obtained they must be reproduced in proper sequence in order to obtain the proper development.

> *Note:* The same analysis would apply for square to circular transition pieces.

Example: In Fig. 8-27 below, the partial plan and front elevation views of the transition piece are given. Since the circle is in the center of the rectangle making both halves symmetrical, it is not necessary to divide the entire circle into a number of equal parts. In fact, in this case only one-quarter of the circle need be divided, but we have divided the near half-portion to further clarify the solution. Connect the numbered points on the circle to the rectangular base points A and D. These lines will form the bases of a series of triangles whose altitude is equal to the vertical height of the transition piece, and whose hypotenuse will be equal to the true distances from either A or D to the points located on the circumference. (See the true length diagram located to the right of the front elevation view.) Assume the transition is to be made in two parts having seams at 5-E and 7-F.

To begin the half development, draw line AD equal in length to the plan view of AD. From both points A and D draw arcs having a radius equal to the true length of A-1 and D-1 as shown in the true length diagram. The intersection of these arcs determines point 1 of the development. Next, with points A and D as centers and radii equal to the true lengths of A-2, A-3, A-4, etc., as shown in the true length diagram, draw arcs of indefinite length. Set the compass to the chordal distance between the numbered points on the near semi-circle in the plan view, and, beginning at point 1 in the development, step off four spaces on either side. Through the intersection points thus obtained draw lines to points A and D as shown. From points 5 and 7 in the development draw arcs equal to the true length of 5-E and 7-F as shown in the front elevation view. Using points A and D in the development as centers, draw arcs equal to A-E and D-F which are shown true length in the plan view. The intersections thus obtained will locate points E and F in the development.

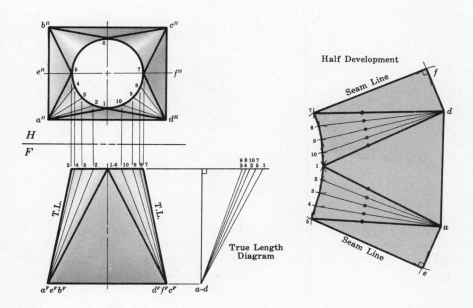

Fig. 8-27

Solved Problems

1. **Given:** The line AB is the axis of a hollow cylinder having a diameter of $1\frac{1}{2}''$ (see Fig. 8-28 below). Point B is $1\frac{1}{2}''$ due east of A and $1\frac{1}{2}''$ below A. Line XY pierces the cylinder. Point X is $1\frac{1}{16}''$ west, $1\frac{1}{4}''$ north of A and $1\frac{1}{2}''$ below A. Point Y is $1\frac{1}{2}''$ east, $\frac{7}{8}''$ south of A and at the same elevation as A. Scale: $\frac{1}{2}'' = 1''$.

Problem: Using the two views only, determine the two points where the line XY pierces the cylinder.

Solution:

Using the given data, draw partial plan and front elevation views of the line and cylinder. Pass a plane parallel to the axis of the cylinder and containing the line XY. This cutting plane cuts across the upper base of the cylinder at points 1 and 2 in the plan view. These two points determine the upper ends of the two straight line elements which lie in the cutting plane. Draw both elements 1 and 2 parallel to the axis in both views. The intersection of elements 1 and 2 with line XY are the required piercing points P_1 and P_2. These pierce points may be checked by simply projecting between views.

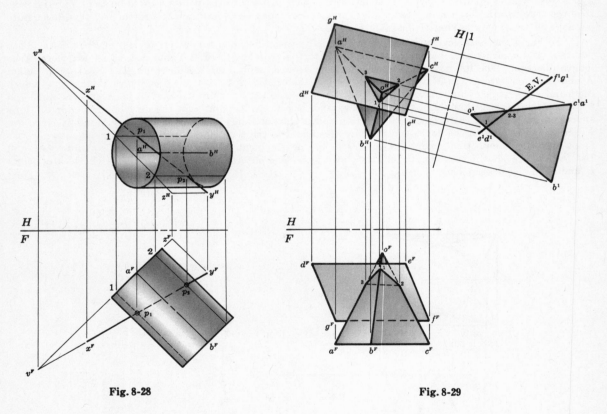

Fig. 8-28   Fig. 8-29

2. **Given:** A right triangular pyramid (base ABC, vertex O) and plane $DEFG$. $A(1\frac{1}{2}, 2, 8\frac{1}{2})$ $B(2\frac{1}{4}, 2, 6\frac{1}{2})$ $C(3\frac{1}{2}, 2, 8)$ $O(2\frac{1}{2}, 4, 7\frac{1}{2})$ $D(1, 3\frac{3}{4}, 7\frac{1}{2})$ $E(3, 3\frac{3}{4}, 7)$ $F(3\frac{1}{2}, 2\frac{1}{2}, 8\frac{1}{2})$ $G(1\frac{1}{2}, 2\frac{1}{2}, 9)$. Refer to Fig. 8-29 above. See Art. 1.7 for the coordinate system of problem layout.

Problem: Find the intersection of the plane and the pyramid. Show complete visibility.

Solution:

Draw the plan and front elevation views of the plane and pyramid. Draw auxiliary view 1 to show the pyramid and edge view of the plane. Label the intersections of the plane with edges OB, OC, and OA of the pyramid 1, 2, and 3 respectively. Project the intersection points to the plan and front elevation views. Careful visualization will determine proper visibility.

3. **Given:** Partial plan and front elevation views of two intersecting prisms in Fig. 8-30. Scale: $\frac{1}{2}'' = 1''$.

Problem: Determine the line of intersection of the two prisms.

Solution:

In the plan view extend the three edges of the level triangular prism until they intersect the edges of the vertical triangular prism. Label the points as shown on the plan view 1 through 6 and A, B, and C. Project these points down to the front elevation view and label them in this view also. Pass a vertical cutting plane coincidental with the edge of the vertical prism where points C, 6, and B are located. Label the intersection of the plane with the top ridge of the level prism, point X. Project point X down to its corresponding position in the front elevation view. Draw a line from point X to point 6 in the front view. The intersection of this line with the vertical edge on which point B is located will determine point 7 which is the required point of intersection common to both planes. Label point 7 in the plan view. Careful visualization will determine correct visibility.

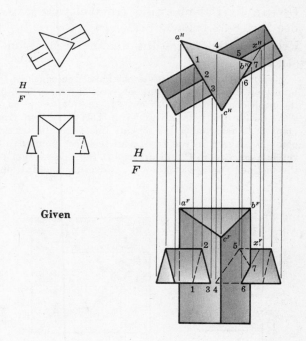

Fig. 8-30

4. **Given:** Partial plan and front elevation views of the intersecting prism and pyramid in Fig. 8-31 below. Scale: $\frac{1}{2}'' = 1''$.

Problem: Determine the line of intersection of the prism and pyramid.

Solution:

Draw the given views. Label the pyramid base $ABCD$ and vertex O, as shown. In the plan view extend the edges of the pyramid until they intersect the vertical prism. Label these intersection points 1 through 4 and project them to the front view. Label intersection points E, F, and G in the plan view. To locate points E, F, and G in the front view, assume vertical cutting planes passing through the plan view of these points. Project the intersection of the cutting plane with the edges of the pyramid down to the front view. The intersection of the cutting plane with the pyramid edges OA, OB, and OD will limit the vertical edges of the prism-points E, F, and G in the front view. Careful visualization will determine proper visibility.

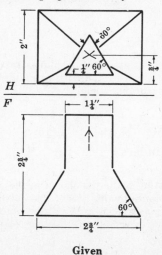

Given

Fig 8-31

5. **Given:** Partial plan and front elevation views of an oblique pyramid and a prism. See Fig. 8-32 below. Scale: $\frac{1}{2}'' = 1''$.

Problem: Determine the line of intersection of the two solids.

Solution:

Draw the two given views. Label the edges of the vertical triangular prism A, B, and C. In the plan view label the apparent intersection points 1, 2, 3, and 4. Locate these points in their corresponding positions in the front view. In the plan view pass a cutting plane coincidental to vertical surface AC. Locate the points 5 and 6 in both plan and front views. In the front view connect point 5 to point 3. The intersection of this line with the vertical edge from A will determine the intersection point D. Again in the front view connect point 6 to point 1. The intersection of this line with the vertical edge from C will determine the intersection point F. The vertical edge from B will intersect the inclined surface at point E. Careful visualization will determine proper visibility.

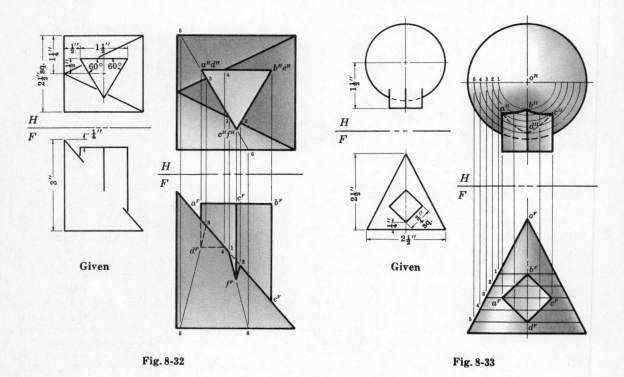

Fig. 8-32   Fig. 8-33

6. **Given:** Partial plan and front elevation views of a prism intersecting a cone in Fig. 8-33 above. Scale: $\frac{1}{2}'' = 1''$.

Problem: Show the intersection of the cone and prism.

Solution:

Draw the given views. In the front view pass a series of horizontal cutting planes through both the prism and cone as shown. In the plan view locate the intersection of the cutting planes with the extended edges of the prism. Careful visualization will determine proper visibility.

7. **Given:** Two intersecting cylinders. See Fig. 8-34 below. Scale: $\frac{1}{2}'' = 1''$.

Problem: Show the line of intersection of the two cylinders.

Solution:

Draw the given views. Pass a series of parallel vertical cutting planes as shown in the plan view. Draw a right section of the inclined cylinder. Show the cutting planes in this view. The cutting planes determine common points which lie on the intersection of the cylinders such as X, Y, 1, and 9. Project these points of intersection to the front view from both the plan view and the inclined right section view. Careful visualization will determine proper visibility.

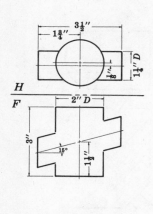

Given

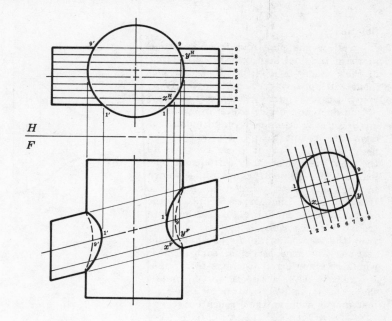

Fig. 8-34

8. **Given:** Two right circular cones are each $2\frac{1}{2}''$ high and have $2''$ diameter bases which are parallel. The vertex of one cone is located $1''$ west and $\frac{1}{2}''$ south of the other vertex. See Fig. 8-35. Scale: $\frac{1}{2}'' = 1''$.

Problem: Determine the intersection of the two cones.

Solution:

Using the given data, draw partial plan and front elevation views of the two right circular cones. In the front elevation view pass two horizontal cutting planes through both cones. In the plan view show the circles cut from each cone by these two cutting planes. The intersection of corresponding circles will locate common points on the line of intersection of the two cones. To locate the highest point of intersection, draw vertical cutting plane AB through vertices X and Y in the plan view. The intersection of straight-line elements XW and YV in the front elevation view determines the highest point of intersection of the two cones. Careful visualization will determine proper visibility.

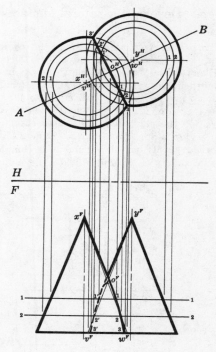

Fig. 8-35

9. **Given:** Plan and front elevation views of the oblique cylinder representing a heating duct which passes through the corner of a factory room. Refer to Fig. 8-36 below. Scale: $\frac{1}{2}'' = 1''$.

Problem: Make a half development of the asbestos necessary to completely cover the duct.

Solution:

Draw the plan and front elevation views of the duct as given. Since the axis of the cylinder is a frontal line, each element will appear true length in the front view. Draw a partial right section view of the duct and divide the elliptical half-circumference into a convenient number of elements, in this case 7, for half of the right section. Project these elements to the front view where they will appear true length. The elements can also be shown in the plan view parallel to the axis. Draw a girth line corresponding to the plane of the partial right section and transfer the spacing between elements from the view showing the one-half right section. The true length of each element would be transferred from the front view. In order to make a smooth curve, connect the element ends as shown by using an instrument such as the French curve.

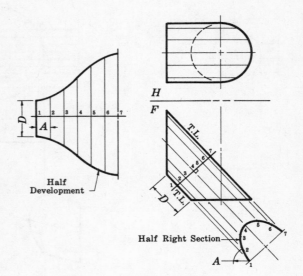

Fig. 8-36

10. **Given:** Plan and front elevation views of the truncated right pyramid shown in Fig. 8-37 below.   Scale: $\frac{1}{2}'' = 1''$.

Problem: Develop the lateral surfaces of the pyramid.

Solution:

Draw the given views. Since the pyramid is a right pyramid having a horizontal truncation, each intersecting edge of the lateral surfaces will be the same length. In the plan view revolve edge C in order to obtain the true length of OC in the front elevation view. Edge CG can now be measured true length in the front view. The true length of each horizontal edge is measured in the plan view. Lay out an arc having a radius equal to the true length of OC. Draw a line from O to C and step off the true length of CG. From point C on the development swing an arc equal to the true length of CD measured on the plan view. The intersection of this arc with the original base arc will locate point D on the development. Connect D to O and step off the true length of DH which is equal to CG. Continue the procedure until the development is complete. An arc from O having radius OG will simplify the location of points H, E, and F in the development.

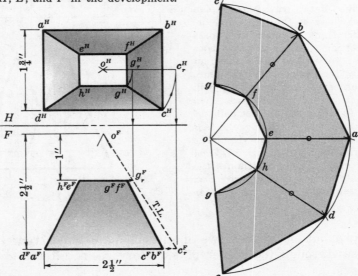

Fig. 8-37

11. Given: Plan and front elevation views of the truncated pyramid.   See Fig. 8-38 below. Scale: $\frac{1}{2}'' = 1''$.

Problem: Develop the lateral surface of the pyramid.

Solution:

Draw the plan and front elevation views as given. Label the lateral edges as shown. Edges *A-1* and *C-3* are shown true length in the front view. To obtain the true length of edges *B-2* and *D-4* revolve point 4 into a frontal plane and measure *A-4* revolved in the front view. Since the true size of the base is seen in the plan view and the true length of the lateral edges are shown in the front view, the development is drawn with point *O* as a vertex center from which the true length edges are constructed. See preceding development.

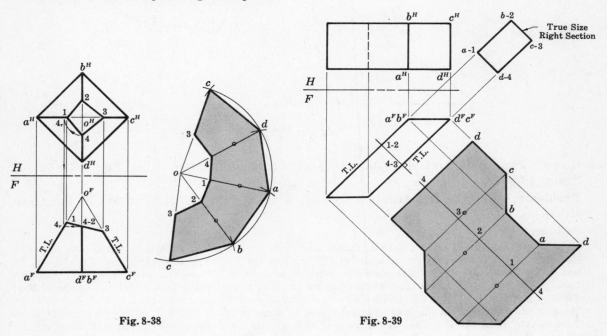

Fig. 8-38   Fig. 8-39

12. Given: Plan and front elevation views of the prism shown in Fig. 8-39 above.   Scale: $\frac{1}{2}'' = 1''$.

Problem: Draw a development of the prism.

Solution:

Draw the plan and front elevation views as given. Label the edges as shown. Draw a view showing the right section in its true size. This right sectional view will show the proper relationship between the lateral edges of the prism. Draw a girth line corresponding to the plane of the right section in the front view. Project the true length of the elements from the front view and the spacing between them from the view showing the true size of the right section. Bend where indicated.

13. Given: An oblique cone has a $1\frac{3}{4}''$ diameter horizontal base.   The vertex is located $2''$ due east of the center of the base and $2\frac{1}{4}''$ above the base.   The cone is to be truncated by a horizontal plane $1\frac{1}{4}''$ above the base.   See Fig. 8-40 below.   Scale: $\frac{1}{2}'' = 1''$.

Problem: Make a half development of the truncated oblique cone.

Solution:

Using the given data, draw the plan and front elevation views of the truncated oblique cone. In the plan view divide the circular base into 12 equal parts, and show the 12 elements from the base to the vertex of the cone. Draw a true length diagram (see Art. 8.16) to determine the true length of each element. With the spacing between the elements and their true length known, the development can now be made by the method of Fig. 8-25.

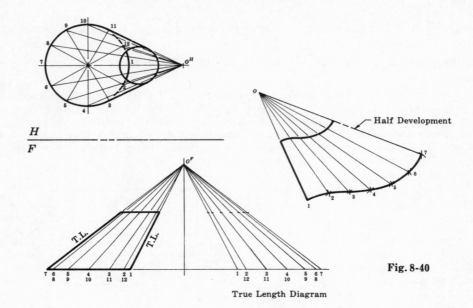

Half Development

True Length Diagram

Fig. 8-40

14. Given: Plan and front elevation views of a transition piece. See Fig. 8-41 below. Scale: $\frac{1}{2}'' = 1''$.

Problem: Using the triangulation method, draw a complete development of the transition.

Solution:

Label each of the eight points as shown and draw a diagonal on each of the four surfaces. To the right of the front elevation view draw XY equal to the height of the transition. From point Y, and to the left, step off the plan view distances of each bend line such as A-E, B-F, etc. Connect these points to point X in the true length diagram in order to obtain the true length of each bend line. To the right of point Y step off the plan view distances of each diagonal such as A-F, F-C, etc. Connect these points to point X in the true length diagram in order to obtain the true length of each diagonal. Assume A-E to be the seam line and draw A-E to begin the development. From point E draw an arc equal to the true length of E-F as shown in the plan view. From point A draw an arc equal to the true length of diagonal A-F as obtained in the true length diagram. The intersection of these two arcs will determine point F in the development. Continue this procedure until the complete development is made. The true lengths of the top and bottom edges of the transition will be obtained from direct transfer of distances in the plan view. The true lengths of the bend lines and diagonals are measured in the true length diagram and these distances are then transferred to the development.

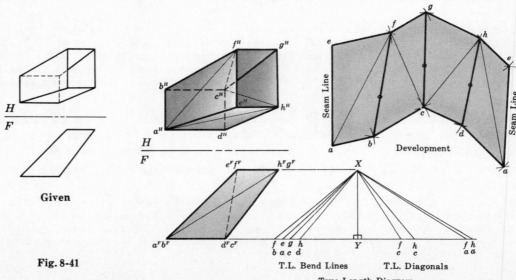

Given

Development

Seam Line

Seam Line

Fig. 8-41

T.L. Bend Lines   T.L. Diagonals

True Length Diagram

15. Given: Partial plan and front elevation views of the transition piece shown in Fig. 8-42 below.   Scale: $\frac{1}{2}'' = 1''$.

Problem: Draw a half development of the transition using the principles of triangulation.

Solution:

Label the points as shown and draw the primary bend lines *A-E*, *B-E*, etc. in both views. Divide the semi-circle into six equal parts and connect these points to the transition base corners *C* and *D*. To the right of the front elevation view draw *XY* equal to the height of the transition piece. From point *X* step off the plan view distance of each bend line such as *D-1*, *D-2*, *D-3*, etc. Connect these points to *Y* in order to obtain the true length of each bend line. Assuming this to be a two-part development, the seam lines would be *G-1* and *F-7*.

Begin the development by drawing a line equal to the true length of *C-D* shown in the plan view. From point *C* draw an arc equal to the true length of *C-4* obtained in the true length diagram. From point *D* draw an arc equal to the true length of *D-4* obtained in the true length diagram. The intersection of these two arcs will locate point 4 in the development. From point *C* draw arcs equal to the true lengths of *C-5*, *C-6*, and *C-7*. From point *D* draw arcs equal to the true lengths of *D-1*, *D-2*, and *D-3*. Beginning at point 4 draw arcs in each direction having a radius equal to the chordal distance between successive circumferential points in the plan view. The intersection of these arcs with the previously constructed arcs from points *C* and *D* will determine points 1 through 7 in the development. From points *C* and *D* draw arcs equal to the plan view distances of *C-F* and *D-G*. Draw an arc from point 7 equal to the vertical height of the transition. The intersection of this arc with the previous arc from *C* will determine point *F* in the development. Draw an arc from point 1 equal to the true length of *G-1* obtained in the true length diagram. The intersection of this arc with the previous arc from point *D* will locate point *G* in the development. Connect the points as shown.

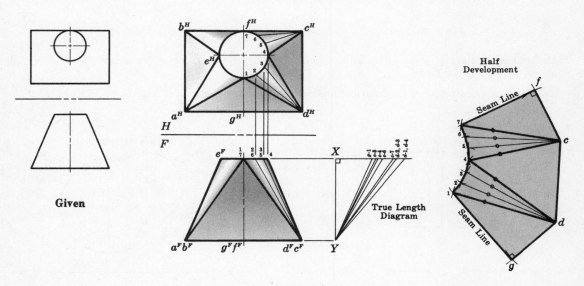

Fig. 8-42

Supplementary Problems

16. Fig. 8-43 below shows the partial plan and front elevation views of two intersecting prisms. Show the intersection of the two prisms. Scale: $12'' = 1'-0''$.

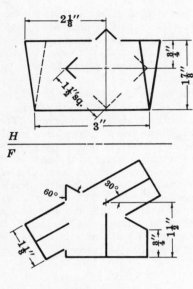

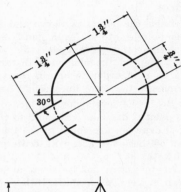

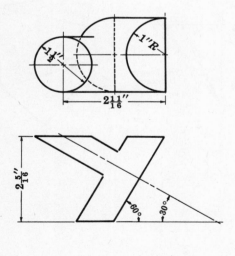

Fig. 8-43

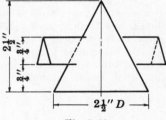

Fig. 8-44

17. Determine the lines of intersection of the right circular cone and triangular prism shown in Fig. 8-44 above. Scale: $12'' = 1'-0''$.

18. Show the intersection of the cylindrical and semi-cylindrical ducts. See Fig. 8-45. Scale: $12'' = 1'-0''$.

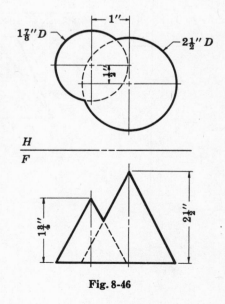

Fig. 8-45

Fig. 8-46

19. Fig. 8-46 above shows the partial plan and front elevation views of two right circular cones which intersect. Determine the line of intersection of the two cones. Develop the largest cone. Scale: $12'' = 1'-0''$.

20. Draw the given views and develop the lateral surfaces of each prism shown in Fig. 8-47 below. Scale to suit.

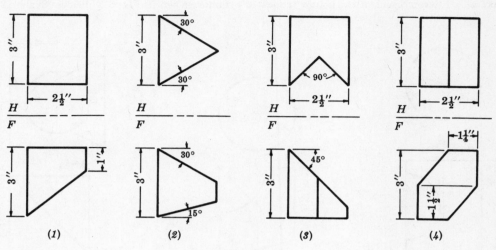

(1)   (2)   (3)   (4)

Fig. 8-47

21. Fig. 8-48 shows the plan and front elevation views of a truncated right pyramid. Make a complete development of the pyramid. The pyramid is a transitional duct. Scale: $12'' = 1'-0''$.

22. Draw the given views and develop the lateral surfaces of the oblique pyramid. Refer to Fig. 8-49 below. Scale: $12'' = 1'-0''$.

23. Fig. 8-50 below shows the plan and front elevation views of a hollow cylinder. Make a half development of the cylinder. Scale: $12'' = 1'-0''$.

24. Fig. 8-51 below shows the plan and front elevation views of a truncated right cylindrical sleeve. Make a half development directly off the front elevation view. Scale: $12'' = 1'-0''$.

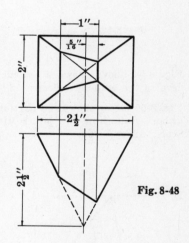

Fig. 8-48

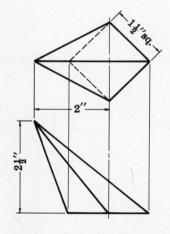

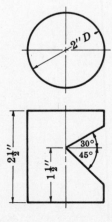

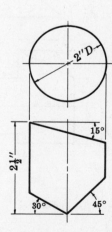

Fig. 8-49   Fig. 8-50   Fig. 8-51

25. Fig. 8-52 below shows a cylindrical solid. Draw the given views and develop the lateral surface of the cylinder. Scale to suit.

26. Make a half development of the hollow truncated cylinder shown in Fig. 8-53 below. Scale: $12'' = 1'-0''$.

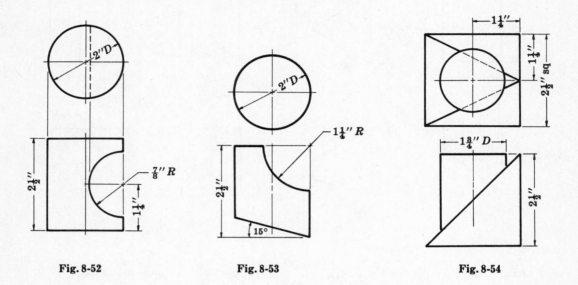

Fig. 8-52   Fig. 8-53   Fig. 8-54

27. Fig. 8-54 above shows a cylinder intersecting an oblique pyramid. Determine the intersection, and develop the cylinder. Scale: $12'' = 1'-0''$.

28. Fig. 8-55 below shows the partial plan and front elevation views of a horizontal cylinder and vertical prism. Determine the intersection of the cylinder and prism. Make a development of each piece. Scale: $12'' = 1'-0''$.

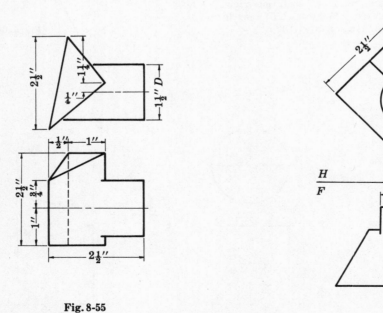

Fig. 8-55   Fig. 8-56

29. Fig. 8-56 above shows the partial plan and front elevation views of a cylinder and prism. Show the intersection of the cylinder and prism. Draw a half development of the cylinder. Scale: $12'' = 1'-0''$.

30. Show the intersection of the two circular cylinders in Fig. 8-57 below. Draw the development of each cylinder. Scale: $12'' = 1'-0''$.

31. Fig. 8-58 below shows partial plan and front elevation views of two intersecting circular cylinders. Determine the intersection of the two cylinders and make half developments of each. Scale: $12'' = 1'-0''$.

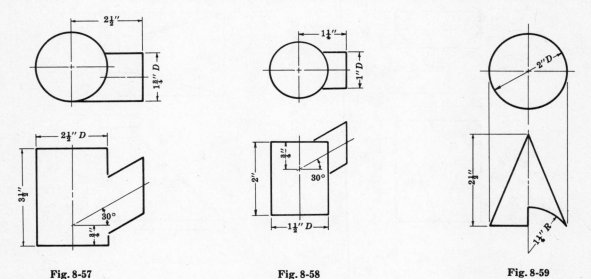

Fig. 8-57   Fig. 8-58   Fig. 8-59

32. Complete the plan view and develop the hollow right circular cone shown in Fig. 8-59 above. Scale: $12'' = 1'-0''$.

33. Draw the given views and make a half development of the lateral surface of the oblique cone shown in Fig. 8-60 below. Scale: $12'' = 1'-0''$.

34. Fig. 8-61 below shows the front and partial plan views of a right circular cone intersected by a triangular prism. Determine the intersection of the prism and cone. Make complete developments of both the prism and cone. Scale: $12'' = 1'-0''$.

35. Fig. 8-62 below shows the plan and partial front elevation views of a truncated cone and intersecting cylinder. Determine the intersection of the cone and cylinder. Make half developments of each. Scale: $12'' = 1'-0''$.

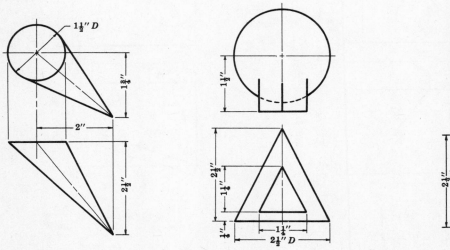
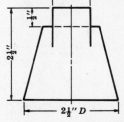

Fig. 8-60   Fig. 8-61   Fig. 8-62

36. Fig. 8-63 below shows the plan and front elevation views of a square to rectangular transition piece. By means of triangulation, draw a half development. Scale: $12'' = 1'-0''$.

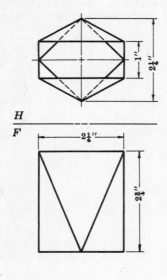

Fig. 8-63

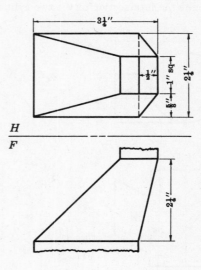

Fig. 8-64

37. Fig. 8-64 above shows the plan and front elevation views of a square to rectangular transition piece. Draw a half development using the method of triangulation. Scale: $12'' = 1'-0''$.

38. Fig. 8-65 below shows the plan and front elevation views of a cylindrical to rectangular transition piece. Draw a complete development of the transition piece using the triangulation method. Scale: $12'' = 1'-0''$.

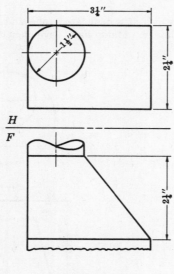

Fig. 8-65

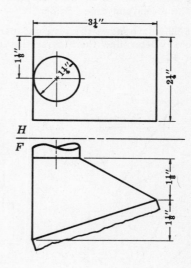

Fig. 8-66

39. Fig. 8-66 above shows the plan and front elevation views of a cylindrical to rectangular transition piece. By means of triangulation, draw a half development. Scale: $12'' = 1'-0''$.

Chapter 9

Mining and Topographic Problems

The basic relationships of points, lines, and planes which have been explained thus far find their practical application in the fundamental problems encountered by students in Mining Engineering and topographic works. For these students there are several texts which adequately cover the more involved problems encountered in this field of endeavor. The purpose of this brief chapter is, therefore, merely to acquaint the student of Engineering or Science with the elementary relationship of Descriptive Geometry to Mining and Topography.

Before a student can attempt the solution of mining problems, he must understand the meaning of terms used in mining operations. The following terms and their meanings are essential for a clear understanding of the basic mining problems (see Fig. 9-1).

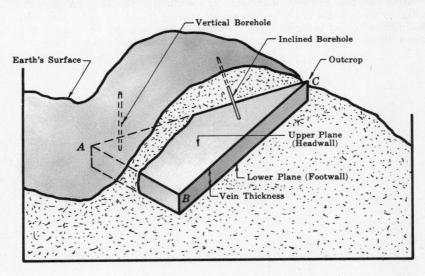

Fig. 9-1. Portion of a Stratum

9.1 DEFINITIONS

(1) *Stratum*—a layer, vein, or seam of ore, usually considered to be an inclined plane of uniform thickness. The *thickness* of the stratum is the distance between the parallel surfaces of the stratum.

(2) *Headwall*—the upper bedding plane, or top surface, of the stratum.

(3) *Footwall*—the lower bedding plane, or bottom surface, of the stratum.

(4) *Borehole*—the hole drilled from the earth's surface to the stratum in order to determine the position of the stratum.

(5) *Outcrop*—the location of the intersection of the stratum with the earth's surface.

(6) *Strike*—the bearing of a horizontal line on the plane of the stratum.

(7) *Dip* — the true slope angle of the stratum, measured at right angles to the strike and downward from a horizontal plane.

(8) *Shaft* — an opening, generally vertical, from the earth's surface to the mining activities.

(9) *Tunnel* — an underground passageway, level or inclined, from the shaft to the stratum.

Note: There are several other terms necessarily used in Mining but the above are sufficient for the basic problems covered in this chapter.

9.2 To DETERMINE the STRIKE and DIP from GIVEN MAP DATA

Analysis: Sufficient data is given to locate points on the stratum. Locate a level line on the plane and show its bearing, or strike, on the plan view. Since the dip is the same as slope angle, any elevation view showing the plane as an edge will also show the dip of the vein.

Example: The given map data is shown in Fig. 9-2(*a*) below. The elevation of point *C* is 350′. This data is transferred to the plan and front elevation views of Fig. 9-2(*b*). A level line is drawn on the front view of the plane from point *C*. The true length of this line, as shown in the plan view, will determine the strike of the plane as being N 70° W.

The auxiliary elevation view 1 shows the plane of the vein as an edge, and the slope angle of 45° is the dip of the stratum. The dip of a stratum is usually represented on a map by drawing a short line away from the strike line and perpendicular to it. This dip line should point toward the low portion of the plane. The dip, or slope angle of 45°, is shown by placing the angle value alongside the dip line. The dip is then verbally described as being 45° SW. A stratum having a dip of 45° SW is quite different from a stratum having a dip of 45° NE, even though the strike in both cases is the same.

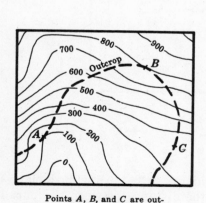

Points *A*, *B*, and *C* are outcrop points on the upper plane of a stratum.

(*a*) Given Map Data

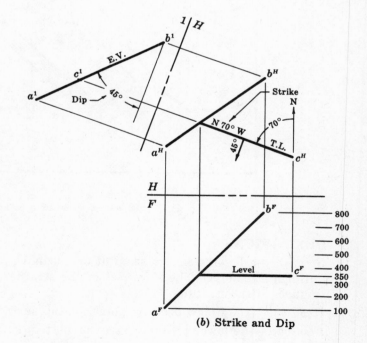

(*b*) Strike and Dip

Fig. 9-2. Determining the Strike and Dip

9.3 To DETERMINE the STRIKE, DIP, OUTCROP, and THICKNESS of a VEIN

Analysis: Three points determine a plane. A level line shown in the plan view will determine the strike of the vein. The dip, being the same as the slope angle, will be measured in an elevation view which shows the plane as an edge. A point on the footwall is located by a vertical borehole intersecting the lower surface of the stratum. Assuming the vein of ore to be of uniform thickness and therefore parallel, an edge view of the footwall is drawn parallel to the headwall, or upper plane of the stratum. The thickness of the vein is determined by the perpendicular distance between the two plane surfaces. An elevation view showing the vein of ore as an edge will also show the earth's irregular contour lines as straight horizontal lines. This elevation view will show the intersection of each contour line with the vein of ore. Projecting these points of intersection to the plan view will determine the plan view of the outcrop.

Example: The contour map of a portion of mining property is shown in Fig. 9-3 below, with points A, B, C, and D given. The points A, B, and C are three points on the upper plane of the vein. B and C are points of outcrop on the upper plane. Point D is located on the lower plane of the vein and directly below point A.

The strike is determined by drawing a level line on the front view of plane ABC and showing the bearing of this level line in the plan view. An auxiliary elevation view, projected from the plan view and showing the plane as an edge, will determine the dip. Assuming the upper and lower planes to be parallel, we may draw the lower plane by locating point D in view 1. The thickness of the vein would be the perpendicular distance between the lines representing the upper and lower bedding planes.

The intersection of plane ABC with the 260' level in the auxiliary elevation view determines points X and Y, which are located on both the vein and the earth's surface. These two points X and Y, when projected back to the plan view, will determine two points on the upper outcrop line. Additional points on the upper and lower outcrop lines are determined in the same manner. All the visible ore is located within the upper and lower outcrop lines.

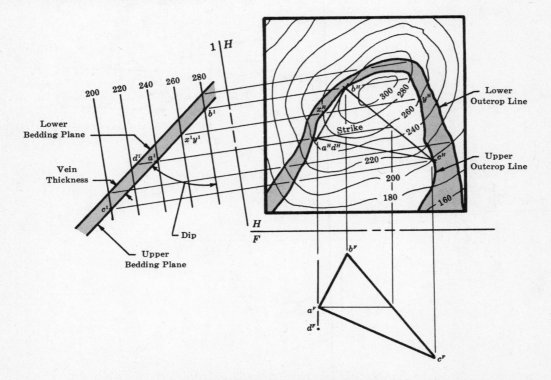

Fig. 9-3. Strike, Dip, Thickness, and Line of Outcrop of a Vein

9.4   To DETERMINE the STRIKE, DIP, and THICKNESS of a VEIN USING TWO NON-PARALLEL BOREHOLES

Analysis: Two non-parallel boreholes will intersect the headwall and footwall at two points on each surface.

A.   **Line Method:**   The two points on the headwall should be connected with a straight line. A view which shows this line as a point will also show the headwall as an edge even though its direction is as yet unknown.   Since the headwall and footwall are assumed to be parallel, the view will also show the footwall as an edge, and its position is, therefore, determined by locating the two points on the footwall surface.   The headwall can now be drawn parallel to the footwall and the vein thickness will be apparent.

   The plan view of a level line on the headwall will determine the strike of the vein.   An elevation view showing the headwall and footwall as parallel edges will also show the dip of the vein.

   Note: The student should be certain that the dip is measured in an elevation view and not in an inclined view.

   Example: In Fig. 9-4 below, a vertical borehole from *A* intersects the headwall at *B* and the footwall at *C*. An inclined borehole from *D* intersects the headwall at *E* and the footwall at *F*. The points *B* and *E* on the headwall are connected and this line is shown in its true length in the auxiliary elevation view 1. The point view of line *BE* will appear on the edge view of the headwall in the inclined view 2. The points *C* and *F* on the footwall will appear in the inclined view 2, and since both the headwall and footwall are assumed to be parallel, a line drawn through *C* and *F* will also determine the direction of the headwall through the point view of line *BE*. The thickness of the vein is measured in this view.

   To determine the strike of the vein, a level line of random length is drawn on the headwall in view 1 and projected to the headwall shown as an edge in view 2. This level line can now be located in the plan view to determine the strike. A new auxiliary elevation view showing the strike as a point will also show the headwall and footwall as parallel lines, confirming the vein thickness obtained in view 2 as well as giving the dip of the vein.

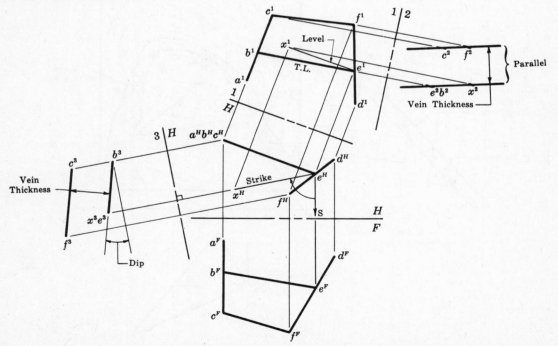

**Fig. 9-4.   Strike, Dip, and Thickness of a Vein Using 2 Non-Parallel Boreholes
(Line Method)**

B. **Plane Method:** Connect the two points on the headwall and also connect the two points on the footwall. Using the method of Article 3.1, draw a plane containing one of the lines and parallel to the other line. Draw an elevation view showing the edge view of the plane, parallel to the other line. Since this elevation view shows both the headwall and footwall as edges and parallel to each other, both the dip and thickness of the vein can be measured in this view. The strike of the vein would be determined by the plan view of a level line drawn on the headwall or footwall.

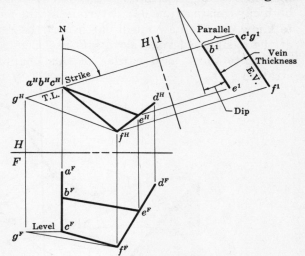

Example: See Fig. 9-5. Having the same given data as in the line method, draw a plane containing line *CF* and having line *GF* parallel to line *BE* in both plan and front elevation views. Make line *GF* of sufficient length in the front view so that *CG* will be a level line. Thus it will appear in true length in the plan view and will determine the strike of the vein. The auxiliary elevation view 1 will show the plane *CFG* as an edge parallel to the headwall. Both the dip and vein thickness are measured in this view.

Fig. 9-5.
Strike, Dip, and Thickness of a Vein
Using 2 Non-Parallel Boreholes
(Plane Method)

9.5 GEOLOGIST'S COMPRESSED METHOD for FINDING the APPARENT DIP WHEN the TRUE DIP and STRIKE ARE KNOWN

One of the most frequent problems encountered by Geology students is that of determining the apparent dip of a stratum when the true dip and strike are known. By definition, the apparent dip of a stratum is the slope angle of any line in the plane of the stratum other than the dip line. As we have already learned, the true dip is the slope angle of the stratum measured perpendicular to the strike and downward from a horizontal plane. An apparent dip line, not being perpendicular to the strike line, must always have a slope angle smaller than the true dip.

Analysis: Draw a line parallel to the given strike line and intersecting the plan view of both the given true dip line and the apparent dip line. This new strike line can be drawn at any distance from the given strike line, but its location will determine the vertical distance it lies below the given strike. If the true dip line is revolved 90° into the horizontal projection plane, the true dip angle will appear. If the apparent dip line is revolved 90° into the horizontal projection plane, the apparent dip will appear and will measure less than the true dip.

Example: In the adjacent Fig. 9-6, the strike *AB* and the true dip angle are given. An auxiliary strike line is drawn any convenient distance from the given strike line, intersecting the plan view of both the true dip line and apparent dip line at points *C* and *E*, respectively. The true dip is constructed at *A* (angle *CAD*), thus determining point *D*. The line *EF* is drawn perpendicular to the plan view of the apparent dip line, *BE*, and is equal in length to the line *CD*. The apparent dip angle is *EBF*.

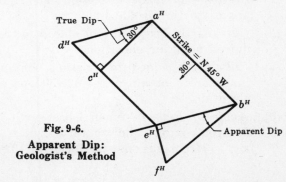

Fig. 9-6.
Apparent Dip:
Geologist's Method

9.6 CUTS and FILLS

Of the many types of problems encountered by civil engineers, one of the most common is that of determining the extent of earth cuts and fills necessary in railroad and highway construction.   In determining the location of a new highway it is vitally important to know exactly how much earth must be removed (cut) from a hillside in order to fill in at the designated road elevation (see Fig. 9-7 below).   The principles involved are quite similar to those used earlier in this chapter in the determination of the outcrop of a stratum.

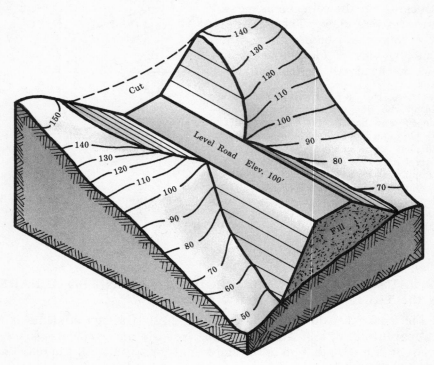

Fig. 9-7

The following are some of the terms used in locating cuts and fills:

(1) *Profile* — a vertical section of the earth's surface containing a given line which may be either straight or curved. The length of the profile must be equal to the true length of the given line. (See Fig. 9-8.)

(2) *Section* — a vertical section at right angles to the profile line.

(3) *Cut* — earth removed to obtain a required slope or elevation.

(4) *Fill* — earth added to existing contour in order to obtain a required slope or elevation.

(5) *Angle of Repose* — when loose earth is dumped onto a horizontal surface a right circular cone is formed.   The

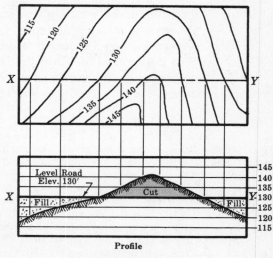

Fig. 9-8

maximum angle of slope that the side of the earth, or other such material, can have without further sliding is called the angle of repose. This slope is usually designated by a ratio whose first number is the horizontal distance and whose second number is the vertical distance.

Example: Fig. 9-9 below shows a contour map and the proposed location of a level road at 100′ elevation and of a given width. The angle of repose for cuts and fills is $1\frac{1}{2}$:1. It is required to show the limits of both cuts and fills.

Draw a typical section of the road, showing the parallel contour intervals for both cuts and fills. Draw lines having the correct angle of repose from both edges of the road. On the contour map draw cut and fill contour lines parallel to the centerline of the road and projected from the section. The intersection of these parallel cut and fill contour lines with their corresponding natural contour lines will determine points on the required cut and fill lines. Connect the points as shown to determine the exact limits for both cuts and fills.

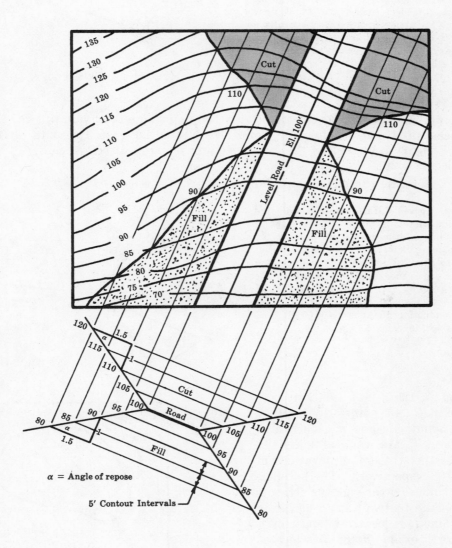

Fig. 9-9

Solved Problems

1. **Given:** *A*, *B*, and *C* are three points on the upper bedding plane of a stratum. Point *B* is 110′ east, 230′ north and 120′ above *A*. Point *C* is 270′ east, 80′ north and 100′ above *A*. See Fig. 9-10. Scale: 1″ = 300′.

   Problem: Determine the strike and dip of the stratum.

   Solution:
   Using the given data locate the three points in the plan and front elevation views. Connect the three points to represent a plane in both views. A level line in the front view will determine the strike in the plan view. Place folding line *H*-1 perpendicular to the strike and draw auxiliary elevation view 1 to show the plane as an edge. The dip is measured in this view.

   Ans. Strike = N 58° W, Dip = 25°

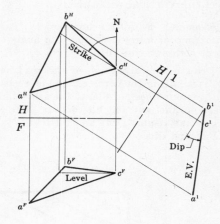

Fig. 9-10

2. **Given:** Points *A*, *B*, and *C* are on the upper bedding plane of a vein of ore. Point *B* is 10′ due east of *A* and 7′ below *A*. Point *C* is located 20′ east, 10′ south of *A* and 8′ above *A*. Point *D* is located on the lower bedding plane 8′ directly below *C*. See Fig. 9-11. Scale: 1″ = 20′.

   Problem: Determine the strike, dip, and thickness of the vein.

   Solution:
   Using the given data, draw the plan and front elevation views of plane *ABC* and point *D*. A level line on the plane will reveal the strike in the plan view. Draw auxiliary elevation view 1 to show plane *ABC* as an edge, and point *D*. The dip is measured in this view. The lower bedding plane can be drawn parallel to the edge view of plane *ABC* and through point *D*. The perpendicular distance between the two parallel lines is the vein thickness.

   Ans. Strike = N 72° W
   Dip = 66°
   Thickness = 3′–3″

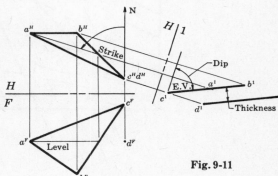

Fig. 9-11

3. **Given:** Fig. 9-12(*a*) is a contour map showing point *C* as a point of outcrop on the upper bedding plane of a stratum. Points *A* and *B* are vertical boreholes which strike the stratum at elevations of 160′ and 210′ respectively. The borehole from *B* continues on through the stratum and reaches the footwall at an elevation of 195′. Point *B* is located 65′ east of *A* and 115′ north of *A*. Point *C* is located 140′ east of *A* and 85′ north of *A*. Scale: 1″ = 100′.

   Problem: Determine the strike, dip, and thickness of the stratum.

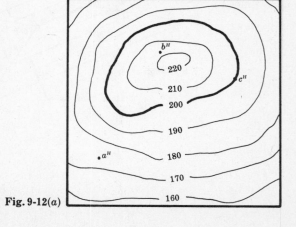

Fig. 9-12(*a*)

Solution:

Draw the plan and front elevation views, showing points *A*, *B*, and *C* on the upper bedding plane, and locate the point *D* on the lower bedding plane as shown in Fig. 9-12(*b*). A level line on the upper bedding plane will determine the strike of the stratum in the plan view. Draw auxiliary elevation view 1 to show the plane *ABC* as an edge. The true dip of the stratum is measured in this view. Locate point *D* in this view and draw a line through *D* parallel to the edge view of plane *ABC*. The perpendicular distance between the two bedding planes determines the thickness of the stratum.

Ans. Strike = N 86° W
    Dip = 22°30′
    Thickness = 13′−6″

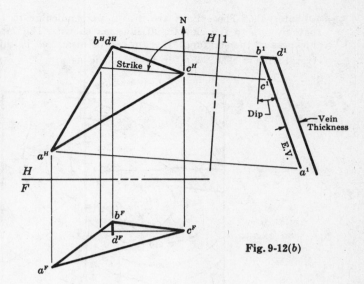

Fig. 9-12(*b*)

4. **Given:** The upper bedding plane of an ore stratum containing points *A* and *B* has a strike of N 60° W and a dip of 45° SW. $A(2, 1\frac{1}{2}, X)$ $B(4\frac{1}{2}, 2\frac{1}{4}, 5)$. See Fig. 9-13. Scale: 1″ = 40′.

Problem: Determine the true distance from *A* to *B*.

Solution:

Using the given data locate points *A* and *B* in the front view and point *B* in the plan view. Construct a plane *ABC* in the front view having line *BC* level. Line *BC* in the plan will have the given strike of N 60° W. Locate folding line *H*-1 perpendicular to the strike line and draw the point view of line *BC*. Since view 1 is an elevation view, the 45° dip of the plane can be shown in the view. Locate point *A* in view 1 by transferring the distance from the front view. Project from view 1 to locate *A* in the plan view. By revolution locate the revolved position of point *A* in the plan, and measure the true distance from *A* to *B* in the plan view.

Ans. True Distance = 53′−9″

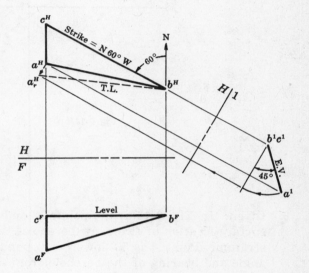

Fig. 9-13

5. **Given:** Fig. 9-14(*a*) shows a contour map whose shaded area reveals the outcrop of a stratum. Scale: 1″ = 60′.

Problem: Determine the strike, dip, and thickness of the stratum.

Solution:

Connect two points of the same outcrop line and having the same elevation as shown in Fig. 9-14(*b*) below. Here points *A* and *B* were chosen on the 240′ level line of the contour map. The angle between line *AB* and due north will determine the strike. Off to the side of the contour map draw parallel elevation lines having

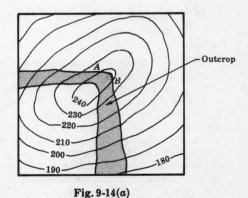

Fig. 9-14(*a*)

10′ intervals. These lines are drawn perpendicular to the strike. Project several points from the contour map to the elevational lines as shown. The intersections thus obtained will determine the thickness of the stratum. The dip can be measured in this view also.

Ans. Strike = N 55° W, Dip = 45°, Thickness = 13′

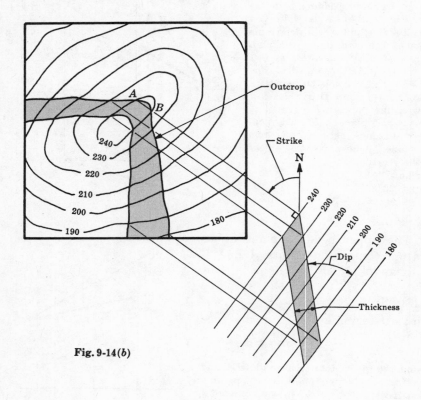

Fig. 9-14(*b*)

6. **Given:** In Fig. 9-15(*a*) the points *A* and *B* indicate the surface location of two skew boreholes that penetrate a stratum. Point *A* is 50′ lower than point *B*. The slope angle and bearing of the borehole from *A* are shown in the figure. From *A* the headwall and footwall are reached after drilling 75′ and 300′ respectively. From *B* the headwall and footwall are reached after drilling 50′ and 125′ respectively. Scale: 1″ = 200′.

Problem: Determine the strike, dip, and thickness of the stratum.

Solution:

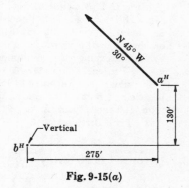

Fig. 9-15(*a*)

Draw the given view as shown in Fig. 9-15(*b*) below. Locate folding line *H*-1 parallel to the bearing of the borehole from *A*. Draw an auxiliary elevation view 1 showing the slope of the borehole from *A* as well as points *W* and *X* representing the headwall and footwall respectively. Project *W* and *X* to the plan view. Show the borehole *AX* in the front view. Draw the front elevation view of the vertical borehole from *B* showing *Y* and *Z* representing the headwall and footwall respectively. Connect *Y* to *W* and *Z* to *X* in both plan and front elevation views. In the front view construct a plane *YWV* having *YV* level and *VW* parallel to *ZX*. Show plane *YWV* in the plan view again with *VW* parallel to *ZX*. The strike is measured in the plan view. Place folding line *H*-2 perpendicular to the strike and draw auxiliary elevation view 2 showing headwall and footwall parallel to each other. In view 2 the perpendicular distance is the vein thickness. The dip is also measured in this view.

Ans. Strike = N 83° E, Dip = 15°, Thickness = 72′

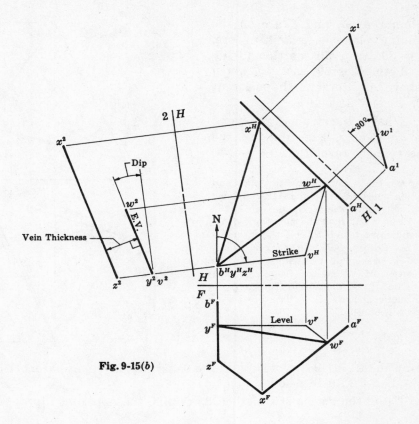

Fig. 9-15(b)

7. Given: Points *A*, *B*, and *C* are on a stratum of ore (see Fig. 9-16). A vertical borehole is to be drilled from point *X* on the surface of the ground. *B* is 110′ east, 40′ south and 20′ above *A*. Point *C* is 80′ east, 60′ north and 70′ above *A*. Point *X* is located 60′ due east of *A* and 80′ above *A*. Scale: $1'' = 60'$.

Problem: Using two views only, determine how deep the hole must be drilled in order to reach the stratum. What is the strike of the plane?

Solution:

Using the given data draw the plan and front elevation views of the plane *ABC* and point *X*. Draw a level line in the front view and measure the strike in the plan view. In the plan view draw a line from point *A* through point *X* until it intersects *BC* at *D*. Show the line in the front view of the plane. In the front view draw a vertical line from point *X* to the line *AD*. Label the intersection point *Y*. Measure *XY* in the front view to obtain the required length of the vertical borehole.

Ans. Borehole = 55′−6″
        Strike = N 56°30′ W

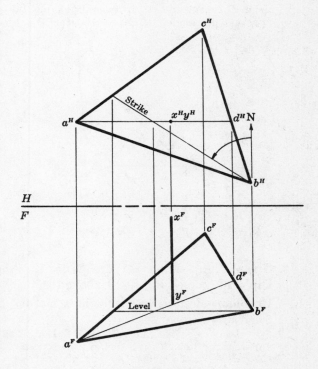

Fig. 9-16

8. **Given:** Points *A*, *B*, and *C* are on the upper bedding plane of a vein of ore. Point *B* is 30′ east, 60′ north and 45′ below *A*. Point *C* is 60′ east, 15′ north and 20′ above *A*. Refer to the adjacent Fig. 9-17. Scale: 1″ = 40′.

Problem: What is the strike of the vein? Using two views only, determine the true dip of the vein by revolution.

Solution:

Using the given data draw the plan and front elevation views of the plane *ABC*. A level line *AD* will determine the strike as seen in the plan view. From point *B* in the plan view draw a line perpendicular to *AD*. Label the intersection point *E*. Locate *BE* in the front view. Revolve *BE* in the plan view and show the revolved position of *BE* in the front elevation view. The true dip is measured from a horizontal base line as shown.

Ans. Strike = N 60°30′ E
          Dip = 50°

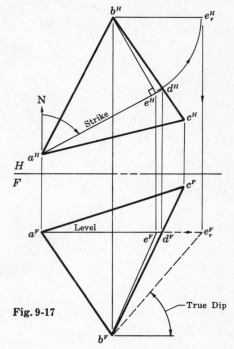

Fig. 9-17

9. **Given:** Map data: Strike = S 45° W, True Dip = 50° NW, Apparent Dip Direction = Due West. See Fig. 9-18 below.

Problem: Using the geologist's method determine the apparent dip.

Solution:

Draw a strike line from point *A* at S 45° W. From point *A* draw the true dip at 50° NW. Construct a line parallel to the strike. Draw the apparent dip line *DE* in a due west direction. From point *E* construct a right triangle having the distance *EF* = *BC*. The angle *EDF* is the required apparent dip angle.   *Ans.* Apparent Dip = 40°

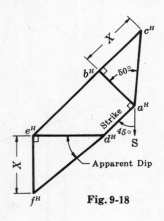

Fig. 9-18

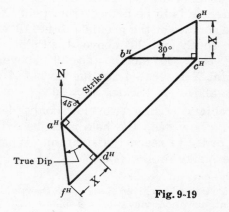

Fig. 9-19

10. **Given:** Map data: Strike = N 45° E, Apparent Dip Direction = Due East, Apparent Dip = 30°. See Fig. 9-19 above.

Problem: Using the Geologist's method determine the true dip.

Solution:

Draw a strike line from point *A* at N 45° E. Terminate the strike line at *B* and draw the apparent dip line *BC* in a due east direction. Construct the apparent dip angle of 30°. Draw a line *CD* parallel to the strike. At point *C* construct a perpendicular line terminating at *E*. Draw line *AD* perpendicular to the strike line, and construct a right triangle having *DF* equal to *CE*. The angle *DAF* is the required true dip angle.   *Ans.* True Dip = 39° SE

11. Given: Fig. 9-20(*a*) below shows a contour map having *XY* representing the center line of a proposed level highway. The road is 20′ wide and is at an elevation of 100′. The angle of repose for both cuts and fills is 1:1. Scale: 1″ = 40′.

Problem: Determine the extent of the cut and fill lines.

Solution:

Using the given data draw the 20′ wide highway with *XY* as the center (see Fig. 9-20(*b*) below). Draw a typical section of the road showing the parallel contour lines at 5′ intervals with the angle of repose at 1:1. Project the cut and fill contour lines from the section back to the contour map, and draw them parallel to the center line of the road. The intersection of these parallel cut and fill lines with their corresponding natural contour lines will determine points on the lines of cut and fill. Connect these points as shown to locate the exact limits of both cuts and fills.

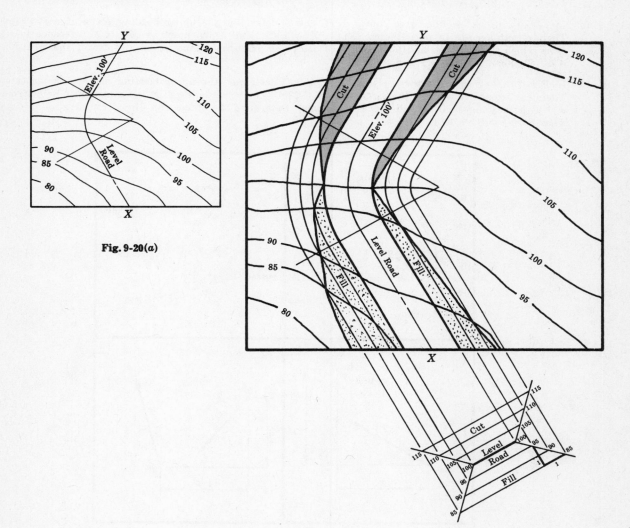

Fig. 9-20(*a*)

Fig. 9-20(*b*)

Supplementary Problems

In each of the following problems determine and label the strike and dip of the plane *ABC*. See Art. 1.7 for the coordinate system of problem layout.

			Strike	Dip
12.	$A(1, 1, 4)$ $B(2, \frac{1}{2}, 3\frac{1}{2})$ $C(3\frac{1}{2}, 2, 4\frac{1}{2})$		*Ans.* S 83°30′ E	52°
13.	$A(1\frac{1}{2}, 1\frac{1}{4}, 4)$ $B(2\frac{1}{2}, 2, 4\frac{1}{2})$ $C(3\frac{1}{4}, \frac{1}{2}, 3\frac{1}{2})$		*Ans.* Due East-West	56°30′
14.	$A(1\frac{1}{4}, 2, 5)$ $B(2\frac{1}{4}, 3, 4)$ $C(3, 3\frac{1}{2}, 5\frac{1}{4})$		*Ans.* N 6° E	41°30′
15.	$A(4, 1\frac{1}{2}, 5)$ $B(4\frac{1}{2}, 1, 3)$ $C(6, 1\frac{1}{2}, 4\frac{1}{2})$		*Ans.* S 76° E	15°30′
16.	$A(3\frac{1}{2}, 1\frac{1}{4}, 4\frac{1}{2})$ $B(5, 1, 3\frac{1}{4})$ $C(7, 1\frac{1}{4}, 3\frac{1}{2})$		*Ans.* S 74° E	17°30′
17.	$A(2, 2, 7)$ $B(3, 3\frac{1}{2}, 5)$ $C(4\frac{1}{2}, 2\frac{1}{2}, 6\frac{1}{2})$		*Ans.* N 85°30′ E	36°

18. In Fig. 9-21(*a*) below, the three points *A*, *B*, and *C* determine the upper bedding plane of a stratum. Their elevations are in parentheses. Scale: $1'' = 100'$. The bench mark (B.M.) is a reference point from which the bearings and map distances are given. Determine the strike and dip of the stratum. *Ans.* Strike = N 54°30′ E, Dip = 29°

19. In Fig. 9-21(*b*) below, the three points *A*, *B*, and *C* determine the upper bedding plane of a stratum. Their elevations are in parentheses. Scale: $1'' = 200'$. The bench mark (B.M.) is a reference point from which the bearings and map distances are given. Determine the strike and dip of the stratum. *Ans.* Strike = S 56°30′ E, Dip = 21°30′

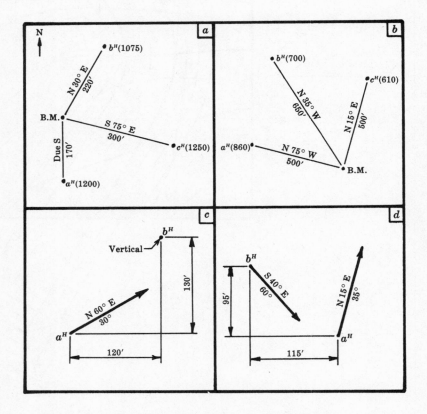

Fig. 9-21. Problems Involving Strike, Dip, and Thickness of Veins

20. In Fig. 9-21(*c*) above, the points *A* and *B* indicate the surface location of two skew boreholes that penetrate a vein of ore. The slope angle and the bearing of the borehole from *A* is shown in the figure. Point *A* is located 40′ lower than point *B*. From *A* the headwall and footwall are reached after drilling 30′ and 100′, respectively. From *B* the headwall and footwall are reached after drilling vertically 25′ and 60′, respectively. Scale: $1'' = 50'$. Determine the strike, dip, and thickness of the vein. *Ans.* Strike = N 61° E, Dip = 29°, Thickness = 31′

21. In Fig. 9-21(d) above, the points A and B are surface locations of two non-parallel boreholes that penetrate a vein of sandstone. Point A is 25' above point B. The slope angle and bearing of each borehole
are given in the figure. From A the upper bedding plane and footwall are reached after drilling 50'
and 100', respectively. From B the upper and lower bedding planes are reached after drilling 40' and
125', respectively. Scale: 1'' = 50'. Determine the strike, dip, and thickness of the stratum.
Ans. Strike = N 54°30′ W, Dip = 43°, Thickness = 47'

22. In Fig. 9-22 points X and Y are located on a stratum of ore
which has a strike of N 60° W and a dip of 45° SW. Scale:
1'' = 60'. Locate the plan view of point X. What is the bearing of
a line connecting X and Y?   *Ans.* Bearing = N 88°30′ W

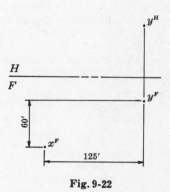

23. Two non-parallel boreholes are drilled toward a stratum of ore.
The borehole from point A has a bearing of N 45° W and has a
slope of 35°. It reaches the headwall of the vein after drilling to
an elevation of 240' below the surface. The footwall is reached
after drilling to an elevation of 360' below the surface. The borehole from B, located 300' due east of A and at the same elevation
as A, has a bearing of N 45° W and has a slope of 45°. It reaches
the headwall of the vein after drilling to an elevation of 220'
below ground, and the footwall is reached after drilling to an
elevation of 430' below ground. Scale: 1'' = 200'. Using the plane
method, determine the strike, dip, and thickness of the stratum.
Ans. Strike = N 79° W, Dip = 72°, Thickness = 54'

Fig. 9-22

24. Using the same given data as for problem 23, determine the strike, dip, and thickness of the stratum
using the line method. Scale: 1'' = 200'.   *Ans.* Strike = N 79° W, Dip = 72°, Thickness = 54'

25. Points A, B, and C are located on the upper bedding plane of an ore stratum (see Fig. 9-23 below).
Point B is also on the outcrop line. Point D is located on the lower bedding plane 50' directly below C.
The boreholes at A and C reach the upper bedding plane of the stratum at elevations of 200' and 300',
respectively. Trace the contour map and determine the strike, dip, and thickness of the vein. Draw
the outcrop lines. Scale: 1'' = 50'.   *Ans.* Strike = N 75°30′ W, Dip = 63°30′, Thickness = 22'

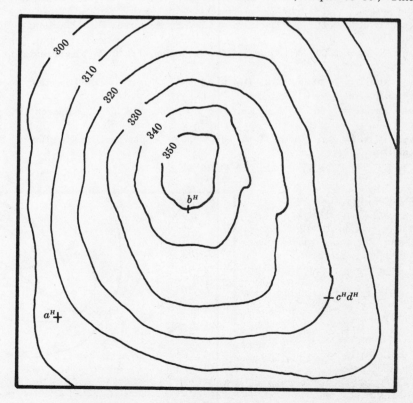

Fig. 9-23

26. A vein of ore is determined by plane *ABC*. Line *XY* is a mine shaft heading in the direction of the stratum. Point *B* is 100′ east, 75′ south of *A* and 85′ above *A*. Point *C* is 200′ east, 50′ north of *A* and 25′ above *A*. Point *X* is 150′ east, 50′ south of *A* and 25′ below *A*. Point *Y* is 250′ east, 90′ south of *A* and 75′ below *A*. Scale: 1″ = 50′. Using two views only, determine how much the shaft must be lengthened in order to reach the vein of ore. Determine the strike and dip of the vein.
Ans. Lengthen shaft 115′, Strike = S 67° W, Dip = 38°30′

27. Points *A*, *B*, and *C* are located on the headwall of a stratum. Point *X* is located on a shaft from which a tunnel is to be driven toward the stratum. Point *B* is located 15′ east, 30′ south of *A* and 15′ above *A*. Point *C* is located 45′ east, 10′ north of *A* and 10′ above *A*. Point *X* is 25′ north, 10′ west of *A* and 30′ above *A*. Scale: 1″ = 20′. Determine the shortest distance, bearing, and slope of a tunnel from *X* to the stratum. Show the tunnel in all views.
Ans. Shortest Distance = 37′–9″, Bearing = S 42° E, Slope = 66°

28. Using the same given data as for Problem 27, determine the shortest distance from point *X* to the stratum if the tunnel is to have a grade of 20%. Scale: 1″ = 20′. Determine the strike and dip of the stratum. Show the tunnel in all views.
Ans. Shortest Distance = 65′–6″, Strike = N 51° E, Dip = 24°30′

29. *A* and *B* are points on the upper outcrop line of a vein of ore. A vertical borehole from *C* reaches the upper plane at a depth of 30′ and then reaches the footwall at a depth of 45′. Point *B* is located 50′ west, 250′ south of *A* and 100′ below *A*. Point *C* is located 150′ east, 75′ south of *A* and 70′ below *A*. Scale: 1″ = 100′. Determine the strike, dip, and thickness of the stratum.
Ans. Strike = N 49° E, Dip = 32°, Thickness = 12′

30. Using the same given data as for Problem 29, determine the strike, dip, and thickness of the stratum if the borehole from *C* reaches the headwall and footwall at depths of 45′ and 75′, respectively. Scale: 1″ = 100′.   *Ans.* Strike = N 44° E, Dip = 36°, Thickness = 25′

31. Using the same given data as for Problem 29, determine the strike, dip, and thickness of the stratum if the borehole from *C* reaches the headwall and footwall at depths of 35′ and 60′, respectively. Scale: 1″ = 100′.   *Ans.* Strike = N 47° E, Dip = 34°30′, Thickness = 23′

32. In each of the following problems certain map data is given. Using the geologist's method determine the missing data.
   (*a*) Given: Strike = N 60° W, True Dip = 45° NE, Apparent Dip Direction = Due East. Find: Apparent Dip.
   (*b*) Given: Strike = N 30° W, Apparent Dip Direction = S 40° W, Apparent Dip = 45°. Find: True Dip.
   (*c*) Given: Strike = N 30° E, True Dip = 45° SW, Apparent Dip Direction = N 30° W. Find: Apparent Dip.
   (*d*) Given: Strike = N 60° E, Apparent Dip Direction = S 75° E, Apparent Dip = 35°. Find: True Dip.
   (*e*) Given: Strike = Due North, True Dip = 30°, Apparent Dip Direction = N 30° E. Find: Apparent Dip.
   Ans. (*a*) 26°30′, (*b*) 47° SW, (*c*) 41°, (*d*) 44°30′ SE, (*e*) 16°

33. Trace Fig. 9-24 and draw a profile along centerline *XY* which represents a level highway at 130′ elevation. Designate cut and fill areas with contrasting shade lines. Scale: 1″ = 40′.

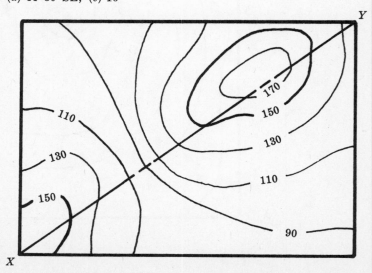

Fig. 9-24

34. Fig. 9-25 below shows a contour map having contour intervals of 10′. It is required to locate a curved highway having its center line in an arc of 150′ radius from point X. The highway is 30′ wide and at a constant elevation of 260′. Using a slope of $1\frac{1}{2}$:1 for both cuts and fills, draw the lines of cuts and fills. Scale: 1″ = 50′.

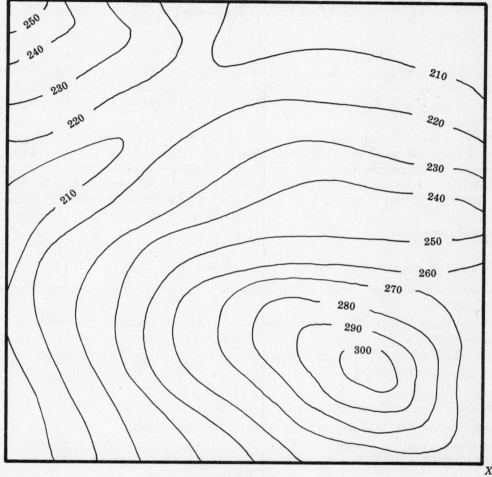

Fig. 9-25

35. Using the same given data as shown above, construct a profile along the center line of the highway.

Chapter 10

Vector Geometry

Physical quantities may be divided into two groups — scalar quantities and vector quantities. Scalar quantities such as temperature, volume, and time have only magnitude, while vector quantities such as force, velocity, and acceleration have both magnitude and direction.

Problems dealing with vectors are often encountered in engineering, and two methods are readily available for their solutions: one deals with mathematics, the other with graphics. The advantages of graphic solutions of vector problems are especially noteworthy, and they make the method a valuable one — complex calculations are eliminated, considerable time is often saved, and a graphic solution makes the problem more easily understood. Too, the accuracy afforded by this type of solution is quite sufficient for that demanded by most vector problems.

10.1 DEFINITIONS
(1) *Vector* — line segment representing a vector quantity. As such, it has a given direction and length.
(2) *Line of Action* — line along which the vector lies coincident. The vector "acts" along this direction.
(3) *Concurrent Vectors* — vectors whose lines of action intersect at a common point.
(4) *Nonconcurrent Vectors* — vectors whose lines of action do not intersect at a common point.
(5) *Coplanar Vectors* — vectors lying in the same plane.
(6) *Noncoplanar Vectors* — vectors *not* lying in the same plane.
(7) *Resultant* — vector obtained by adding two or more vectors. It may replace the vectors added together, as it has the exact same effect as the vectors.
(8) *Equilibrant* — vector needed to exactly balance one or more vectors, in order to maintain equilibrium. It coincides with the resultant of the given vectors in both magnitude and position, but its direction is exactly opposite.
(9) *Space Diagram* — diagram not drawn to scale, showing only the direction and position of the vectors.
(10) *Vector Diagram* — diagram showing the addition or resolution of vectors, with vectors drawn in true direction and magnitude.
(11) *Tension* — force tending to pull an object apart or stretch it.
(12) *Compression* — force tending to squeeze the object together.

10.2 BASIC VECTOR PRINCIPLES
(1) The transmissibility principle states that a vector may act anywhere along its line of action without changing the effect produced by the vector. (See Solved Problem 2)
(2) Two concurrent or parallel vectors are always coplanar, but three or more vectors may or may not be so.
(3) Each vector in the vector diagram must be parallel to its corresponding vector in the same view of the space diagram.

(4) The vector diagram for several forces in equilibrium about a point must be a closed figure, with vectors laid end-to-end in continuous direction. This means the tip of the last vector added must just touch the tail of the initial vector, when the vectors are added by the polygon method, tails-to-tips. This insures that all forces cancel out, leaving no effective force acting at the point; i.e., the system is indeed in equilibrium.

(5) For problems dealing with equilibrium about a point, the following notation is adopted: a vector placed in the member pointing *away* from the point implies a tension in that member; and a vector placed in the member pointing *toward* the point implies a compression in that member.

10.3 RESULTANT of CONCURRENT COPLANAR VECTORS

Two methods are commonly used: the parallelogram method and the polygon method. The latter requires fewer construction lines and is generally preferred. Solutions employing both methods are included in the examples below.

A. Parallelogram Method

Analysis: The resultant of any two vectors, tails together, is the diagonal of a parallelogram constructed with the two vectors as sides.

Example: In Fig. 10-1(*a*) below we are given the space diagram of three forces ($A = 50\#$, $B = 180\#$, $C = 90\#$) and asked to find the resultant.

In Fig. 10-1(*b*) below, step off the right length for each vector (determined by its magnitude) on the line of action of that vector from the intersection point Q, using a convenient scale. Note that A and B form two sides of a parallelogram; complete it by drawing a line parallel to B through the tip of A, and a line parallel to A through the tip of B. Draw in the diagonal from point Q; this is the resultant R_1 of A and B. Construct a second parallelogram with R_1 and the remaining vector C as sides, and draw in their resultant, R_2. R_2 is the desired resultant of A, B, and C, and is the one single force that can replace them. *Ans.* $R_2 = 140\#$, directed as shown

The equilibrant of A, B, and C would be represented by a vector lying along the same diagonal as R_2, but exactly opposite in direction. It is the one force that will just balance A, B, and C.

Note: The order in which vectors are added is immaterial, but the vectors must always be placed tails-together for this method.

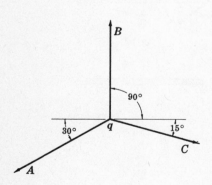

Fig. 10-1(*a*)

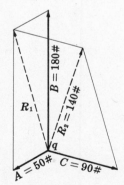

Fig. 10-1(*b*)
Resultant by Parallelogram Method

Fig. 10-1(*c*)
Resultant by Polygon Method

B. Polygon Method

Analysis: Draw in the final side of the polygon formed by adding vectors one after another, the tail of each succeeding vector attached to the tip of the last vector added. This final side, drawn from the tail of the first vector to the tip of the last vector added, is the resultant.

Example: Same problem as above. In Fig. 10-1(c) above, a convenient scale chosen, draw in A to scale from a point Q. Now draw B (direction remaining unchanged) with its tail to the tip of A, and similarly attach C to B. The resultant R is drawn from the tail of A to the tip of C, and is in that direction. This gives us both the magnitude and direction of the resultant.   *Ans. R = 140#*

10.4 RESULTANT of NONCONCURRENT COPLANAR VECTORS

A. Line-of-action Method

Analysis: Move two vectors along their lines of action to the common intersection point, and find the resultant by the parallelogram method. Using this resultant and an unused vector, continue the process until only the final resultant remains.

Example: In Fig. 10-2 we are given three forces ($A = 20\#$, $B = 15\#$, $C = 25\#$) acting on a plank, and asked to find their resultant and its position.

In Fig. 10-3 below, slide A and C (arbitrarily chosen) along their lines of action until they meet at their intersection Q. A scale chosen, mark off the true length of each vector from Q, and find the resultant R_1 by the parallelogram method. Now slide R_1 and the remaining vector, B, along their lines of action until their tails are together at their common intersection P. Drawing each in its true length, use the parallelogram method again to find R_2, the resultant of A, B, and C. We now know the magnitude, position, and direction of the resultant.   *Ans. $R_2 = 50\#$*

This method is good for lines of action that intersect sharply. But for parallel or nearly parallel lines of action, intersecting outside the limits of the drawing, a more general method is needed to determine the position of the resultant.

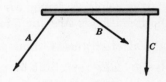

Fig. 10-2

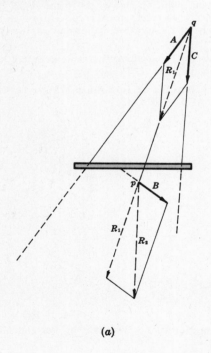

(a)

Fig. 10-3

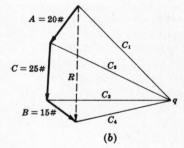

(b)

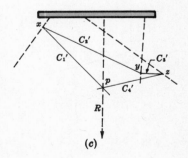

(c)

B. Component Polygon Method

Analysis: The magnitude and direction of the resultant is found by simply using the polygon method in a vector diagram. To find its position, draw construction lines from the tip and tail of each vector in the vector diagram to a point Q off to the side. For construction lines touching two original vectors, draw parallel construction lines between the lines of action of those two vectors in the space diagram, forming part of a polygon. For the two construction lines touching only one original vector in the vector diagram, close the polygon by drawing parallel construction lines from the respective lines of action of those vectors in the space diagram. The intersection of these last two lines is a point on the line of action of the resultant.

Example: Same problem as above. In Fig. 10-3(b) above, draw the vector diagram (polygon method) and construction lines to point Q. Note that C_2 and C_3 touch two vectors. C_2 touches A and C; so, parallel to C_2 draw C_2' in the space diagram (c) between the lines of action of A and C from any point X on the line of action of A. Draw C_3' on a similar basis from the newly found point Y. Since C_1 touches only A and C_4 touches only B, from X draw C_1' parallel to C_1, and from point Z draw C_4' parallel to C_4. Their intersection P is on the line of action of the resultant, because both C_1 and C_4 touch the resultant in (b). Compare the position of the resultant with that found using the other method. *Ans.* $R = 50\#$

10.5 RESULTANT of CONCURRENT NONCOPLANAR VECTORS

Analysis: Since these vectors do not all lie in the same plane, we need two views to determine their positions. Completing the vector diagram in each view will give two views of the resultant, from which we can find its true length, and hence its magnitude.

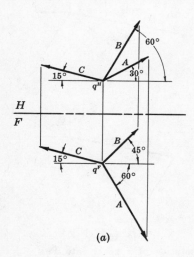

(a)

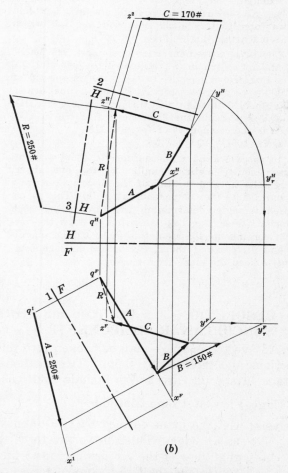

(b)

Example: In Fig. 10-4(a) we are given three force vectors ($A = 250\#$, $B = 150\#$, $C = 170\#$) in a space diagram, and asked to find the resultant.

Before we can complete the vector diagram, we need to get the true length line of each vector in order to mark off its correct magnitude and establish a definite length for it in each view of the vector diagram. Thus, in Fig. 10-4(b), from a point Q, draw a line parallel to A in the two views. Choose a point X on this line, and get the true length line of QX in inclined view 1. On this line mark off 250#, the magnitude of A. Project this distance back to the front and plan views; this gives us the length of A in each view. Using the polygon method, from the tip of A draw

Fig. 10-4. Resultant of 3 Noncoplanar Vectors by Polygon Method

a line parallel to B in both views. In this case, use point Y and revolution to determine the true length line, and mark off 150# thereon. Project this distance back to the front and plan views to get the length of B in both views. From the tip of B now draw a line parallel to C in both views, and get the true length line, using point Z. On this line mark off 170#. Now, with this distance projected back to both views, complete the polygon in each view of the vector diagram by drawing in the dashed line from Q to the tip of C. This is the resultant; draw auxiliary view 3 to find its true length, and thus its magnitude. *Ans.* $R = 250\#$

Note: The parallelogram method could just as well have been used in both views.

10.6 RESOLUTION of a VECTOR into TWO COPLANAR COMPONENTS

Any vector may be broken down into two coplanar components in specified directions, and replaced by them. This process is known as resolution.

Analysis: Form a parallelogram with the single vector as a diagonal, and sides along the component directions. These two sides represent the components of the force in the specified directions.

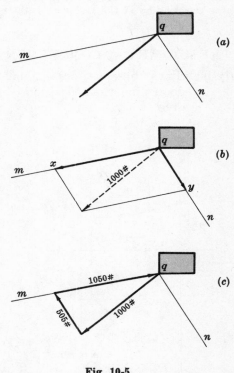

Example: We are given a 1000# force acting on the block as shown in Fig. 10-5(a), and asked to find the two forces acting along lines QM and QN which could replace the single force.

Choose a convenient scale and mark off 1000# along the force line in (b). Extend lines QM and QN until a line drawn parallel to QM through the tip of the 1000# vector intersects QN, and a line drawn parallel to QN through the tip intersects QM. Distance QX represents the component of force in QM, and QY is the component of force in QN.
Ans. $QX = 1050\#$, $QY = 505\#$

In cases dealing with equilibrium about a point, we use another method, similar to the one above. Through the tip and the tail of the force vector draw lines parallel to the components, as in (c). This forms a closed triangle, determining the force along each component. Observing basic principle 4, draw the vectors in continuous direction; this gives us the magnitude and direction of the forces needed in each component to keep the block from moving.

Fig. 10-5
Resolution of a Force into Two Coplanar Components

10.7 RESOLUTION of a VECTOR into THREE CONCURRENT NONCOPLANAR COMPONENTS

Any vector in space may be broken into three specific concurrent components in given directions and replaced by them. Two methods are commonly used, and each has its advantage in special cases. We include solutions to both.

A. Edge-View of Plane Method

Analysis: Get the plane formed by two unknown components — usually forces — to appear as an edge. This allows the force to be resolved into only two components in that view of the vector diagram, which you readily do. One of these components represents the two unknown forces which appear as an edge. Divide this component into the two respective forces with the aid of another view in the vector diagram. Obtain the true length lines of each component to find the actual magnitudes of force. (Note basic principles 3, 4, and 5.)

Example 1: In Fig. 10-6(a), we are given a tripod supporting a 250# vertical load, and asked to find the stresses acting in each leg.

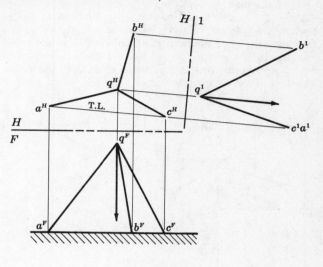

(b)

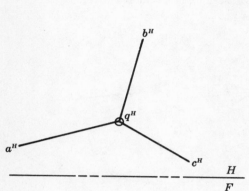

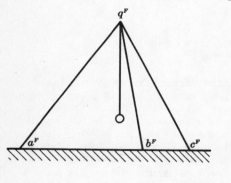

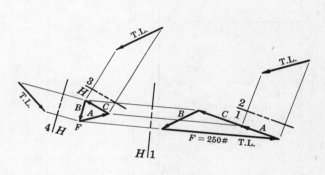

(a)

(c)

Fig. 10-6
**Resolution of a Force into 3 Noncoplanar
Components, Edge-View of Plane Method**

Since line AC is level in the front view of Fig. 10-6(b) above, it is a true length line in the plan view. Draw folding line H-1 perpendicular to AC, making legs QA and QC appear as an edge in the auxiliary elevation view. Now in the vector diagram (c), immediately draw in the known force F in both the plan view and elevation view 1, choosing a convenient scale and marking off 250# in view 1, since the force line appears in true length there. Thus, in view 1 the force may be resolved along the two lines by drawing sides parallel to the components of force in the auxiliary view of (b). Remember that the vector diagram of a system in equilibrium must be a closed figure with the vectors added in continuous direction; hence, the reversed direction of the force components in the legs — they exactly balance the acting force. In the plan view of (c), also determine vector B, the component of force acting along leg QB. The separation point between A and C (components of force acting in legs QA and QC, respectively) is as of yet not known in the auxiliary view 1, but we know A and C lie end-to-end from the tip of F to the tail of B in this view. Hence, in the plan view, draw in lines from the tip of F and the tail of B parallel to the legs QA and QC in that view of (b). Their intersection completes the polygon in that view and determines the desired point of separation of forces A and C in view 1, found by projection. Again, vectors must be drawn in continuous direction in both views. Find the magnitude of the force in each leg by taking views to get true length.

Ans. $A = 105\#$
$\quad B = \ 88\#$
$\quad C = 110\#$

Note that vector B, placed in leg QB, points *toward* the point Q, about which equilibrium was taken. In this manner, note that all stresses are compressions.

Example 2: Solve the above problem, using Bow's notation.

In Fig. 10-7(a) below, having completed the space diagram, note that the vertical force appears as a point in the plan view. Swing it to the side (never between the two components which appear as an edge — QA and QC in this problem) for ease in notation. In the plan view only, place capital letters around the equilibrium point Q, in every space between the force and the frame, and between any two members of the frame itself (Bow's notation should always be used in only one view). Now draw an arrow to indicate which way the forces are to be read around the equilibrium point: the direction is arbitrary. With the arrow drawn as shown, the vertical force reads XT, the force in leg QB reads TV, etc. Had we reversed the direction of the arrow, the force in leg QB would read VT, the vertical force would read TX, etc.

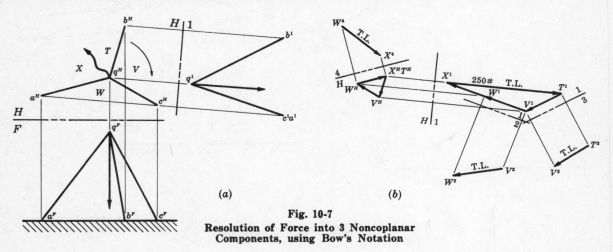

(a)   (b)

Fig. 10-7
Resolution of Force into 3 Noncoplanar
Components, using Bow's Notation

Now in the vector diagram, Fig. 10-7(b) above, draw the only known force first — the vertical load — in both the plan and auxiliary elevation views. Since it reads XT by Bow's notation, label it X at the beginning and T at the end. By the direction of the arrow, vector TV is next. Hence, from T in each view of (b), draw a line parallel to QB of (a). Yet point V is unknown. Note that the remaining vectors VW and WX lie along one straight line in view 1 of (a); so complete the triangle in the elevation view of (b) by drawing the vectors as one straight line from point X, parallel to QAC in the same view of (a). This determines point V, leaving only point W unknown. Now in the plan view of (b), draw a line parallel to WX — plan view (a) — from point X, and a line parallel to VW from point V. Their intersection determines point W, which may be projected to view 1 to determine the length of VW and WX in that view. Draw the vectors in continuous direction, and obtain the true length lines of each vector in order to determine the actual magnitudes of the stresses.   *Ans.* A = 105#   B = 88#   C = 110#

B. Point-View of Line Method

Analysis: Obtain a view in which one unknown component appears as a point. This allows the known force to be resolved into only two components in that view of the vector diagram. As in the previous method, use a related view of the vector diagram to complete the resolution. Derive true lengths to find actual magnitudes.

Example: Same problem as above. Before we can get the point view of a line, we must show the line in true length. Thus, in Fig. 10-8(a), draw auxiliary elevation view 1 to obtain the true length of QB. Place folding line 1-2 perpendicular to QB, getting QB as a point in view 2. The problem is simplified now to resolving the force into two components in this view. In the vector diagram (b) below, establish force F in both views 1 and 2. It appears in true length in view 1, so mark off 250# on this line. Resolve this force into two components in view 2 by drawing lines parallel to the components of view 2 in (a). Hence, in view 1 of (b), draw in vectors A and C — the components of force in legs QA and QC, respectively — by projecting from view 2. Obtain component B by completing the polygon in view 1. Since it appears as a point in view 2, it must be true length in view 1. Vectors must be drawn in continuous direction in both views of the vector diagram. A, B, and C, when placed in their respective members, all point toward the equilibrium point; hence, all stresses are compressions in the legs. Again, the true length lines — as in the previous method — determine the actual magnitudes of A, B, and C.

Ans. A = 105#   B = 88#   C = 110#

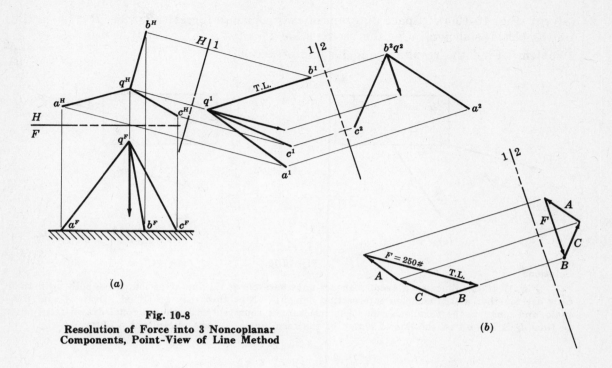

Fig. 10-8
Resolution of Force into 3 Noncoplanar
Components, Point-View of Line Method

Solved Problems

1. **Given:** Fig. 10-9(a). Space diagram showing two coplanar forces ($A = 10\#$, $B = 25\#$) acting at point Q. Use vector scale $1'' = 20\#$.

    Problem: Find the resultant force acting at point Q.

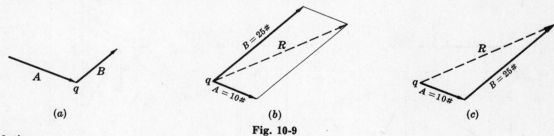

Fig. 10-9

Solution:

   Parallelogram method. See Fig. 10-9(b) above. Slide A along its line of action until its tail is also at Q. Draw A and B to scale, and complete the parallelogram. Draw in the diagonal from point Q; this is the resultant force acting at Q, directed as shown.

   Polygon method. See Fig. 10-9(c) above. Slide A along its line of action until its tail is also at point Q. Mark A to scale, and draw B to scale from the tip of A. Draw a line from the tail of A (point Q) to the tip of B. This is the resultant force, acting at point Q in the direction shown.

   Ans. Resultant $= 31.3\#$.

2. **Given:** Fig. 10-10(a). Space diagram of two coplanar forces ($A = 12\#$, $B = 8\#$) acting on the block as shown. Use the vector scale $1'' = 10\#$.

Problem: Find the resultant and its line of action.

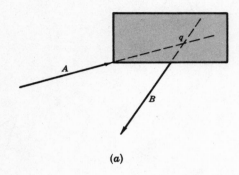

(a)

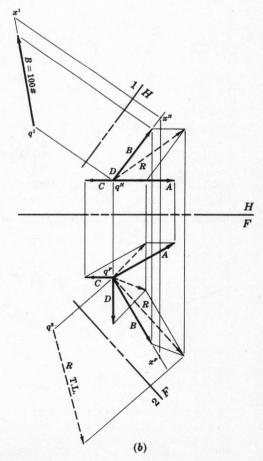

(b)

Fig. 10-10

Solution:

Fig. 10-10(b). Since each vector can act anywhere along its line of action, move both until their tails are together at the common intersection point, Q. Now they may be added. Draw A and B to scale, and complete the parallelogram. The resultant is represented by the diagonal drawn from point Q; thus, Q is a point on the line of action of the resultant.   *Ans.* Resultant $= 7.9\#$

3. **Given:** Fig. 10-11(a). Space diagram showing four noncoplanar forces ($A = 75\#$, $B = 100\#$, $C = 30\#$, $D = 50\#$). Use the vector scale $1'' = 100\#$.

Problem: Find the resultant, using both the parallelogram and polygon methods.

Solution:

Parallelogram method. See Fig. 10-11(b). Slide vectors along their lines of action until their tails are together at the common intersection point, Q. Vectors A, C, and D are shown

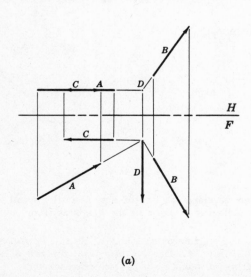

(a)

(b)

Fig. 10-11

true length in the front view, so draw them to scale there. Select a point X on the line of action of B; obtain the true length of QX in auxiliary elevation view 1. On this true length line, mark off 100#, and project back to the other two views to get the length of B in both views. Complete parallelograms in both views until only the final resultant remains. Draw inclined view 2 to get its true length.

Polygon method. See Fig. 10-11 (c). Draw A in both views, marking it to scale in the front view. In both views, draw a line parallel to B from the tip of A; select a point X on this line, and draw auxiliary elevation view 1, obtaining YX in true length. Mark off 100# on this line from Y, and project this distance back to both views. Now, from the tip of B in both views, draw C and mark to scale in either view. Similarly, mark D to scale in the front view, drawing from the tip of C. Draw the resultant from the point Q to the tip of D, and draw inclined view 2 to get its true length.
Ans. Resultant = 125#

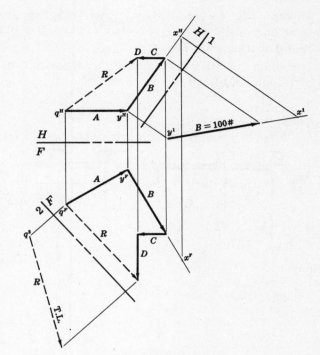

Fig. 10-11(c)

4. **Given:** Fig. 10-12(a) shows a space diagram of three coplanar forces ($A = 70\#$, $B = 90\#$, $C = 60\#$). Use the vector scale $1'' = 100\#$.

Problem: Find the resultant, its vertical and horizontal components, and its line of action by the component polygon method.

Solution:
Fig. 10-12(b). Draw a vector diagram to scale. Find the vertical and horizontal components of the resultant by drawing a vertical line from the tail and a horizontal line from the tip of the resultant. Draw lines from the ends

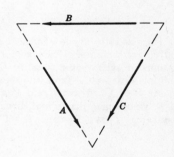

Fig. 10-12(a)

of the vectors to point Q. From a point X on the line of action of B in the space diagram, draw C_2' parallel to C_2. Draw C_3' similarly from the newly found point Y, determining point Z. Find the intersection point, P, of lines C_1' and C_4' drawn from X and Z. P is on the line of action of R.
Ans. Resultant = 142#,   Horizontal comp. = 86#,   Vertical comp. = 113#

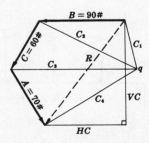

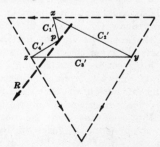

Fig. 10-12(b)

5. Given: Fig. 10-13(a) below. Space diagram showing four weights resting on a weightless bar. ($A = 25\#$, $B = 30\#$, $C = 15\#$, $D = 40\#$). Use the vector scale $1'' = 40\#$ and the distance scale $1'' = 4'$.

Problem: Find the magnitude and position of the single weight which could replace the four, and have the same effect.

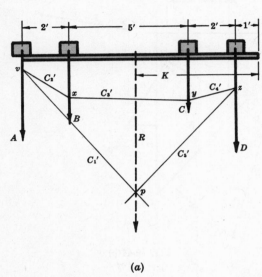

(a)

(b)

Fig. 10-13

Solution:

See Fig. 10-13(b) above. Since the individual forces are parallel, the resultant is simply the sum of the individual weights. Draw vector diagram, with the vectors marked to scale in it. Draw lines from the ends of each vector to point Q. Draw C_2' parallel to C_2 from a point V on the line of action of A. Similarly, draw C_3' from the newly determined point X, and C_4' from point Y. Find the intersection point P of C_1' drawn from V and C_5' drawn from Z. Point P is on the line of action of the resultant R, whose magnitude and direction has already been determined in the vector diagram.
Ans. Resultant $= 110\#$, Distance $K = 5'-3''$

6. Given: Three forces ($A = 600\#$, $B = 800\#$, $C = 750\#$) act on the bridge truss as shown in Fig. 10-14(a) below. Use the vector scale $1'' = 1000\#$ and the distance scale $1'' = 20'-0''$.

Problem: Determine the reactions in the supporting columns.

Solution:

See Fig. 10-14(b) below. Lay off the vectors to scale in a vector diagram, and draw lines from the ends of each vector to point Q. In the space diagram, draw C_2' parallel to C_2 from a point X anywhere on the line of action of A. Similarly, draw C_3' from the newly found point Y. Find the intersection point, P, of C_1' drawn from X and C_4' drawn from Z. Through P, draw the line of action of the resultant, as found in the vector diagram. Now work backwards with the resultant and the horizontal and vertical components of force in the supporting columns. In Fig. 10-14(c) draw the resultant and resolve it into horizontal and vertical components. H represents the combined

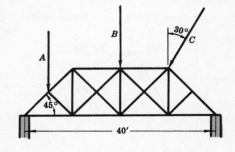

Fig. 10-14(a)

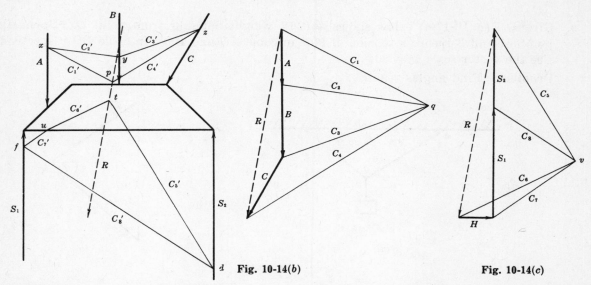

Fig. 10-14(b)   **Fig. 10-14(c)**

horizontal force in the supporting columns. Draw lines C_5, C_6, and C_7 from any point V off to the side as shown. As yet, the desired division of the vertical component into S_1 and S_2 is not known. Arbitrarily choose S_2 to lie above S_1 in (c). In (b), from any point D on S_2, draw C_5' parallel to C_5, between S_2 and the resultant — since C_5 touches these two vectors in (c). From the intersection of C_5' and the resultant, point T, draw C_6' until it meets the line of action of the horizontal force. From this point, U, draw C_7' until it meets the line of action of S_1. Draw a line C_8' from F to D, connecting the two vertical reactions — S_1 and S_2 — in the supporting columns. Determine the division of the total vertical component into S_1 and S_2 by drawing C_8 from V in (c), parallel to C_8' in (b).
Ans. $S_1 = 1180\#$, $S_2 = 870\#$, $H = 375\#$

7. **Given:** An 850# block resting on a 30° inclined slope. See Fig. 10-15. Use the vector diagram $1'' = 1000\#$.

    Problem: Find the force component tending to pull the block down the slope and the perpendicular component pressing the block against the surface.

    Solution:

    Draw an 850# force vector straight down from the center of mass of the block, to scale. Using the parallelogram method, resolve this force into two components, one parallel and one perpendicular to the surface; both pass through the center of mass of the block.   *Ans.* $A = 425\#$, $B = 740\#$

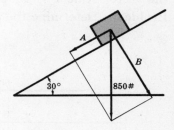

Fig. 10-15

8. **Given:** A 150# weight hanging from the two cables shown in Fig. 10-16(a). Use the vector scale $1'' = 100\#$.

    Problem: Find the tensions in cables QA and QB.

    Solution:

    See Fig. 10-16(b). This is an equilibrium problem. Draw a 150# vertical force vector to scale. From the tip of this vector, draw a line parallel to QA, and a line parallel to QB from the tail; this forms a closed triangle. Draw the vectors in continuous direction. Note that the resultant of A and B exactly balances the 150# force. Placing A and B in their respective members, find that they both point away from the knot; hence, both forces are tensions.   *Ans.* $A = 105\#$, $B = 205\#$

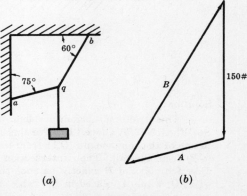

Fig. 10-16

9.　Given:　Fig. 10-17(a) below shows a 120# weight hanging from point Q.　Springs at points A and B record a tension of 90# in cable AQ and 100# in cable BQ respectively. Use the vector scale $1'' = 80\#$.

Problem:　Find angles θ and γ.

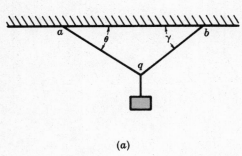

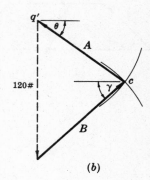

(a)　　　　　　　　　　　　　　　　　　　(b)

Fig. 10-17

Solution:

　　See Fig. 10-17(b) above.　Draw in the 120# vertical force vector from a point Q'.　From the tail of this vector, swing an arc of radius representing 90#; from the tip swing another arc, radius representing 100#.　Draw A and B from the ends of the force vector to the intersection point of the arcs, C. A and B are now parallel to the respective cables QA and QB; hence, measure the angles θ and γ as shown in the vector diagram.　　　*Ans.* $\theta = 35°$, $\gamma = 43°$

10.　Given:　Fig. 10-18(a) below.　A cable and boom holding a 550# ball in suspension.　Use the vector scale $1'' = 400\#$.

Problem:　Determine the stresses acting in both members of the framework.

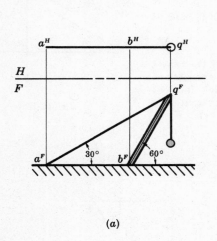

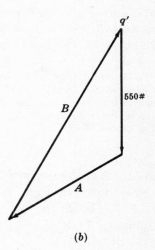

(a)　　　　　　　　　　　　　　　　　　　(b)

Fig. 10-18

Solution:

　　This is an equilibrium problem.　Draw the 550# vertical force vector to scale in a vector diagram. See Fig. 10-18(b) above.　Resolve this force into the two components shown, by drawing a line parallel to one of the components (QA) from the tip of the vector and another line parallel to the other member (QB) from the tail.　Their intersection determines the closed figure shown.　Draw vectors in continuous direction.　A and B exactly balance the 550# force vector.　B, placed in the boom, points toward the point Q; while A, placed in the cable, points away from the point.

Ans. $B = 950\#$, compression, $A = 550\#$, tension

11. Given: A 750# weight is supported by the framework shown in Fig. 10-19 below. Use the vector scale $1'' = 500\#$.

Problem: Find the tension in the cable BC.

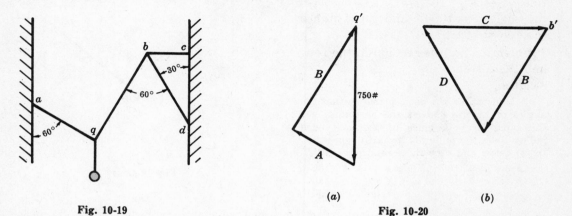

Fig. 10-19   Fig. 10-20

Solution:

First, construct an equilibrium vector diagram about point Q. Draw in Fig. 10-20(a) above the 750# force vector to scale, and resolve the force into two components by drawing lines parallel to QA and QB from the ends of the vector. This determines B, the component of force in QB. Place B in its corresponding member QB, and observe that it points away from the point Q; hence, it is a tension. Now, in Fig. 10-20(b), construct an equilibrium vector diagram about the point B. From a point B', draw in B to scale, directed away from B' since it is a tension. Resolve this known force into two components C and D, acting in the members BC and BD, by the methods used in (a) above. Note that C, placed in member BC, points away from the point B, as indeed a tension should. The combined effect of C and D exactly balances B.   *Ans.* $C = 650\#$ tension

12. Given: A 375# weight hangs at the end of a ten foot weightless boom as shown in Fig. 10-21(a) below. The tension in the cable cannot exceed 1000#. Use the vector scale $1'' = 600\#$ and the distance scale $1'' = 6'-0''$.

Problem: Find the minimum distance X above the boom that the cable may be fastened. What is the stress in the boom when the cable is so fastened?

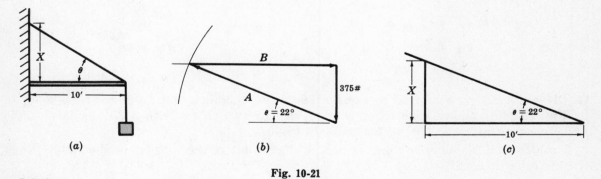

Fig. 10-21

Solution:

We must find the angle θ. In the vector diagram (b) above draw in the 375# vertical force to scale. Resolve this force into two components acting along the given members by swinging an arc from the tip of this force, radius representing 1000#, and finding the intersection point of this arc with a horizontal line drawn from the tail of the vector. A and B act together to exactly balance the force. Measure θ; it equals 22°. Thus, in (c) above construct the triangle, knowing θ; measure X. The stress in the boom is represented by B; placing B in the boom, note that it points toward the equilibrium point. *Ans.* $X = 4'-0''$, $B = 925\#$ compression

13. Given: A 130 ft long rope hangs over a gorge, supporting a 650# weight which has rolled to its natural position. See Fig. 10-22(a). Use the vector scale $1'' = 1000\#$ and the distance scale $1'' = 50'-0''$.

Problem: Find the tension in the rope and the distance X.

Solution:

Understand the basic principle that the angles of the rope above the horizontal are the same on both sides of the suspended pulley when the pulley is in natural position. In (b) below find α by these relations:

$$X \sec \alpha + (100 - X) \sec \alpha = 130$$
$$X \sec \alpha + 100 \sec \alpha - X \sec \alpha = 130$$
$$\sec \alpha = 1.3$$

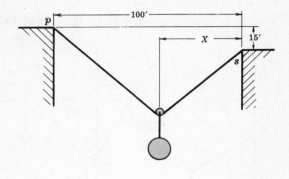

Fig. 10-22(a)

In (c) below construct the angle α at P and S, and find the intersection point Q of the two lines. This determines the distance X. At Q, draw the vertical force to scale, and resolve the force into two components acting along the rope by drawing lines parallel to the rope from both ends of the force vector. Placing these component vectors in the respective members QP and QS, see that they both point away from the point; hence, both feel a tension. Note, too, that tensions in the rope are the same on both sides of the pulley; this will always be true for any rope slipping over a "frictionless pulley".

Ans. Tension = 510#, $X = 42'-6''$

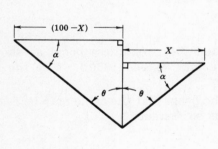

Fig. 10-22(b)

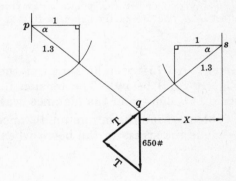

Fig. 10-22(c)

14. Given: A roller is desired to exert a 360# force against the slant surface shown in Fig. 10-23(a) below. The roller is supported by two cables with a 25° included angle between them. Use the vector scale $1'' = 400\#$.

Problem: How heavy a roller is needed, and what is the tension in the cables when this roller is used?

Solution:

See Fig. 10-23(b) below. Draw an equilibrium vector diagram about a point in the center of the roller. Establish the only known force acting *on* the point — a 360# normal force acting *toward* the point, exerted by the slanted surface in order to maintain equilibrium. Resolve this force in (b) into two components, by drawing from the ends of the known vector lines parallel to the other forces acting at the point — the weight of the roller and the tension in the cables. B represents the weight of the roller. To find the actual tension in the cables, take inclined view 1 of A, and resolve A into two components acting along the cables.   *Ans.* Roller = 335#, Cable Tension = 216# in both

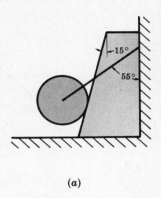

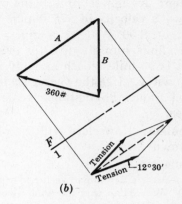

(a)　　　　　　　　　　　　　　　(b)

Fig. 10-23

15. Given: Fig. 10-24. Each cable in the above Problem 14 can stand a maximum tension of 450#. Use the vector scale $1'' = 600\#$.

Problem: Find the maximum roller weight, and maximum force which may be exerted against the surface.

Solution:

Draw a line parallel to the cables in the front view, and draw inclined view 1 of the line. Draw lines at an angle of 12.5° to either side of this line, and on them mark off 450# on each — as they appear in true length in this view. Find the resultant A by the parallelogram method, and project back to the front view to establish A there. Resolve this known force into two equilibrium components parallel to the other forces acting on the center point of the roller — its weight and the normal force of the surface. B represents the maximum roller weight, and C the maximum force against the surface.

Ans. $B = 695\#$, $C = 745\#$

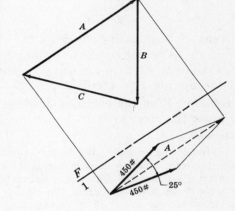

Fig. 10-24

16. Given: A plane originally sets out on a due north heading, but is blown off course by a 75 m.p.h. wind directed S 45° E. The plane shows an air speed reading of 200 m.p.h. Use the vector scale $1'' = 100$ m.p.h.

Problem: Find the actual speed and direction of the plane, both with respect to the earth.

Solution:

Draw a 200 m.p.h. vector directed due north, and attach the wind velocity vector either from its tail or tip. See Fig. 10-25. The actual velocity vector with respect to earth is simply the resultant of the two, found by either (a) the parallelogram or (b) the polygon method.

Ans. Actual Speed = 155 m.p.h.

Actual Direction = N 20° E

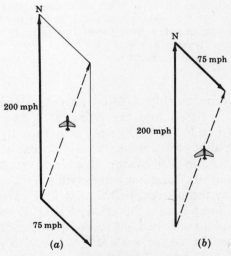

(a)　　　　　　　　　　　　(b)

Fig. 10-25

17. **Given:** A pilot sets his plane on a compass heading of N 20° W; his air speed indicator shows a velocity of 275 m.p.h. At this altitude, a 90 m.p.h. wind is blowing at a direction of S 50° W. At a given time, the pilot finds himself directly over a point that is 90 miles south and 110 miles east of town X. Use the vector scale 1″ = 200 m.p.h. and the distance scale 1″ = 200 miles.

Problem: How close will the plane come to town X?

Solution:

Draw a 275 m.p.h. vector to scale, directed N 20° W. Add to this the wind velocity vector, by either the parallelogram [Fig. 10-26(a)] or polygon method [Fig. 10-26(b)]; their resultant is the actual velocity vector with respect to earth. A perpendicular line drawn from X to the resultant will represent the distance.

Ans. Distance = 28 miles

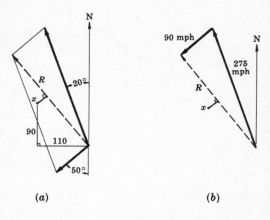

Fig. 10-26

18. **Given:** A boat desires to cross the six mile wide river from point Q directly north to point P. A water current of 4 m.p.h. is heading due east, and the boat itself is capable of a velocity of 6 m.p.h. in still water. Use the vector scale 1″ = 4 m.p.h. and the distance scale 1″ = 4 miles.

Problem: Find the compass heading the boat must set out on. How long will it take to cross the river?

Solution:

Refer to Fig. 10-27. We desire a resultant velocity vector directed straight across the river to point P. Draw the water current vector A to scale from point Q. From the tip of this vector swing an arc, radius representing the 6 m.p.h. velocity of the boat in still water. Find point C, where this arc intersects the straight line drawn from Q to P. The resultant velocity vector is now directed from Q to C when the boat sets out on this compass heading of N θ W. Divide the six mile width by the magnitude of the resultant velocity, in order to find the time elapsed in going from point Q to point P.

Ans. Heading θ = N 42° W

Time = 1 hr. 20 min.

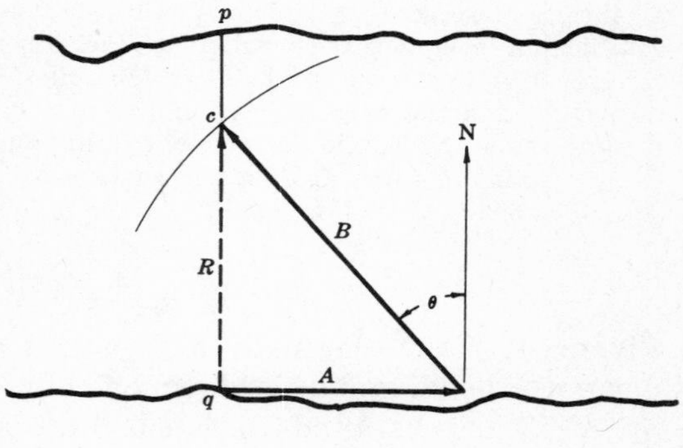

Fig. 10-27

19. **Given:** A ship X is on an actual course of N 15° E, and travelling at a rate of 15 m.p.h. Ship Y, shown in Fig. 10-28, capable of travelling at a velocity of 26 m.p.h. in still water, desires to pass in front of X and not come within 5 miles of it at the nearest point. A water current of 8 m.p.h. in a direction of S 60° E tends to pull the ship Y off its compass heading. Use the vector scale 1″ = 20 m.p.h. and the distance scale 1″ = 20 miles.

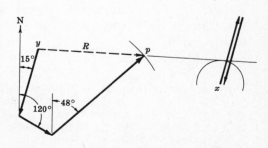

Fig. 10-28

Problem: Find the compass heading of the ship Y if it desires to pass in front of X in the least possible time.

Solution:

In relative motion problems, it is often helpful to introduce another vector acting on both objects, having the effect of setting one of the objects at rest. Set ship X at rest by adding a 15 m.p.h. vector-direction exactly opposite to that of X's motion — to both ships. From the tip of this vector for Y, draw the water current vector; from its tip, swing an arc of radius representing 26 m.p.h. We want the resultant vector to lie along the line from Y tangent to an arc of radius designating 5 miles, drawn about X. Find point P, the intersection of the first arc and the tangent line. This determines both the resultant velocity and the heading of the ship Y.   *Ans.* Heading = N 48° E

20. Given: A plane at A, travelling at an actual speed of 360 m.p.h., is on a course of N 35° E — both with respect to ground (see Fig. 10-29). One minute after this plane passes over A, a second plane leaves B in an attempt to intersect the first plane. This second plane has a velocity of 400 m.p.h. in still air, but a wind velocity of 80 m.p.h. is directed N 15° W. Use the vector scale $1'' = 400$ m.p.h. and the distance scale $1'' = 6$ miles.

Problem: What must be the compass heading of the second plane? Find the time that the second plane is in flight until the intersection point.

Solution:

After one minute, the first plane will be at C. Set the first plane at rest by introducing a 360 m.p.h. velocity vector — directed exactly opposite to its motion — acting on both planes. To the plane at B, add the wind velocity vector, and from the tip of the resultant of the two vectors, swing an arc, radius representing 400 m.p.h. Since we want the resultant of all velocity vectors at B to be directed straight toward C, find the intersection point P of the arc and the line from B to C. This determines both the resultant velocity R and the compass heading of the plane at B. Find the time elapsed by dividing the distance from B to C by this resultant velocity.   *Ans.* Heading = θ = N 3° E, Time = 2.4 minutes

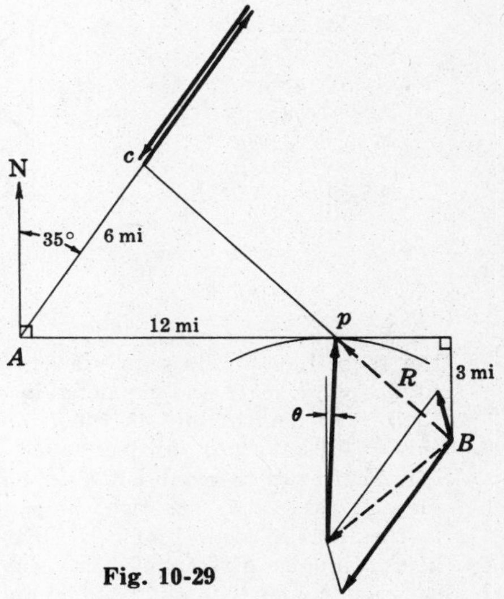

Fig. 10-29

21. Given: The cable and pulley system shown in Fig. 10-30 is pulling up a 300# box at constant velocity. Use the vector scale $1'' = 600\#$.

Problem: Find the stress in each of the three supporting cables, using the edge view of plane method.

Solution:

The tension is the same in all portions of the cable slipping over the pulley. Use this fact to find the *direction* of the resultant force pulling at the intersection point of the three cables in the front view of Fig. 10-31(*a*) below. Draw auxiliary view 1 to get cables QA and QC as an edge. Draw a line parallel to the force F in both views of the vector diagram in Fig. 10-31 (*b*) below. Place an H-F folding line parallel to

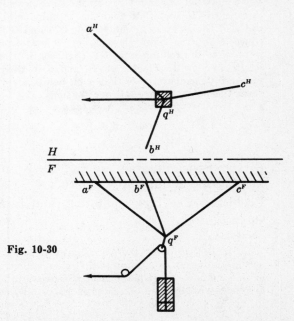

Fig. 10-30

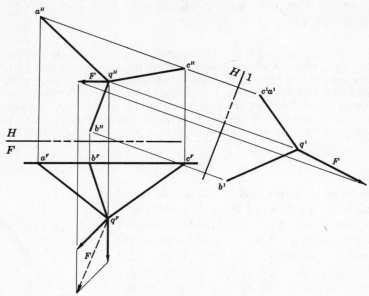

Fig. 10-31(a)

the force line in the plan view — this allows us
to draw to scale the tensions in the cable on
their true length lines in the front view. Find
the resultant and project back to the two
original views to establish F in both views. In
view 1, resolve F into two components parallel
to the lines in view 1 of (a). Draw the vectors
in continuous direction. The separation point
between A and C is as of yet unknown. Project
from view 1 to find B in the plan view. From
the tip of F and the tail of B in the plan view,
draw lines parallel to QA and QC in the plan
view of (a); project the intersection point of
these two lines to view 1 to find the separation
point between A and C. Get the true length
lines of A, B, and C to find the actual magnitudes of the stresses in cables QA, QB, and QC.
Since B in the plan view is, by coincidence,
parallel to the folding line H-1, B shows in true
length in view 1. Placing A, B, and C in their
respective members, note that each points away
from the point. Hence, all are tensions.
Ans. $A = 150\#$, $B = 240\#$, $C = 440\#$

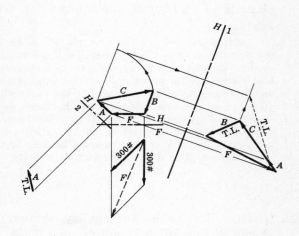

Fig. 10-31(b)

22. **Given:** Space diagram in Fig. 10-32 (a) below showing two cables and a mast, with-
holding the acting force $F = 1000\#$. Use the vector scale $1'' = 1000\#$.

Problem: Use Bow's notation, and find the stresses acting in the mast and in each
cable.

Solution:

   In Fig. 10-32(a), the bent line (for Bow's notation) in the plan view represents the mast, which
actually appears as a point in this view. Hence, it is convenient to resolve the force by the point view
of line method. First place capital letters in all spaces about the equilibrium point D in the plan view,
and choose a direction of reading; this is indicated by the arrow.

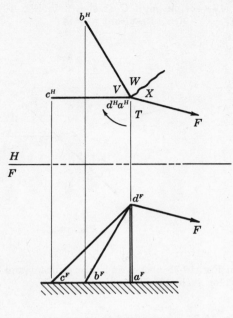

Fig. 10-32(a)

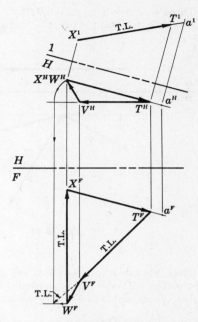

Fig. 10-32(b)

In the vector diagram in Fig. 10-32(b) above establish the known force F as follows: In both views draw lines parallel to F as it appears in the same views of (a). Choose any point A on this line, and obtain the true length line of XA in view 1. Mark off 1000# along this line, and project this distance back to the plan and front views. Since the force is known as XT by Bow's notation, label the beginning of the vector as X and the end as T. Next in order, according to the direction of the arrow, is vector TV. Thus, in both views draw a line parallel to DC as it appears in the corresponding views of (a). Vector VW must complete the triangle in the plan view, because WX appears as a point there. Therefore draw a line parallel to DB — plan view (a) — from point X in the plan view of (b). The intersection of the last two lines drawn determines point V. Project this point to the front view to obtain TV there. Also in the front view of (b), draw vector VW parallel to DB — front view (a) — and project from the plan view to find point W. Complete the polygon in the front view by connecting point W with point X. This vector WX is in true length, as it appears as a point in the plan view. TV is also in true length in the front view, and revolution determines the true length of VW. These true length lines represent the actual magnitudes of the stresses in the various members, drawn to scale. All vectors must be drawn in continuous direction around the figures. Place the vectors in the corresponding members, according to basic principle (5), in order to determine whether the stress is a tension or compression.

Ans. $TV = 1040\#$ Tension, $VW = 370\#$ Tension, $WX = 1210\#$ Compression

Supplementary Problems

23. Two coplanar forces ($A = 5\#$, $B = 12\#$) act on the block shown in Fig. 10-33. Find the equilibrant and resultant. Use the vector scale $1'' = 5\#$.   *Ans.* $E = R = 13\#$

24. Two rockets on a missile are set so that there is an angle of 40° between them; each is capable of giving out a 12,000# thrust. Find the maximum total forward thrust. Use the vector scale $1'' = 5000\#$.   *Ans.* Thrust = 22,500#

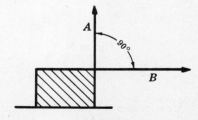

Fig. 10-33

25.　Three concurrent forces ($A = 75\#$, $B = 190\#$, $C = 240\#$) act on point Q as shown in Fig. 10-34 below. Find the resultant.　Use the vector scale $1'' = 100\#$.　　　*Ans.* Resultant $= 373\#$

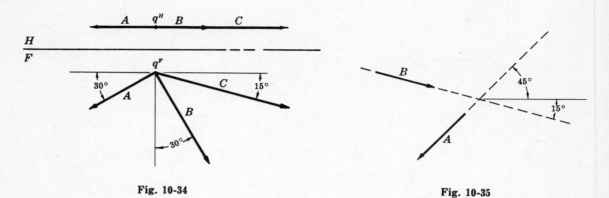

Fig. 10-34　　　　　　　　　　　　　　　　　　　　　　　　Fig. 10-35

26.　Given the two forces $A = 50\#$ and $B = 35\#$ shown in Fig. 10-35 above, find the resultant and the angle it makes with the right horizontal line.　Use the vector scale $1'' = 20\#$.
　　Ans. Resultant $= 44.5\#$, Angle $= 92°$

27.　Five different forces act on the rock as shown in Fig. 10-36 below.　Find the resultant.　Use the vector scale $1'' = 30\#$.　　　*Ans.* Resultant $= 115\#$

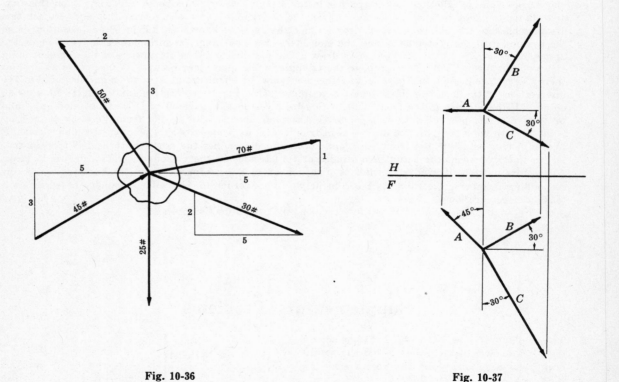

Fig. 10-36　　　　　　　　　　　　　　　　　　　　　　　　Fig. 10-37

28.　Given the three forces $A = 45\#$, $B = 95\#$, and $C = 120\#$ shown in Fig. 10-37 above.　Find the resultant. Use the vector scale $1'' = 50\#$.　　　*Ans.* Resultant $= 95\#$

29. Fig. 10-38 below shows three forces: $A = 300\#$, $B = 520\#$, and $C = 240\#$. Find the resultant. Use the vector scale $1'' = 200\#$.   *Ans.* Resultant $= 584\#$

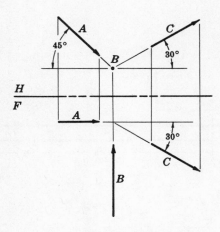

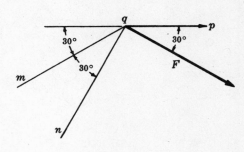

Fig. 10-38   **Fig. 10-39**

30. It is desired to resolve force F ($F = 750\#$) into three coplanar components along lines QM, QN, and QP (see Fig. 10-39 above). The component along QP is known to equal $300\#$ in the direction shown. Find the components of force in QM and QN. Use the vector scale $1'' = 200\#$.
    Ans. $QM = 985\#$, $QN = 1000\#$

31. Force A ($56\#$, $30°$ slope) acts on the block shown in Fig. 10-40 below. Find the component of force pushing the block to the left, in the direction shown. Use the vector scale $1'' = 20\#$.
    Ans. Force $= 42\#$

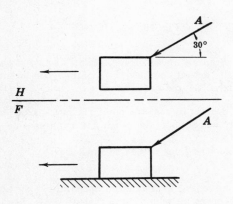

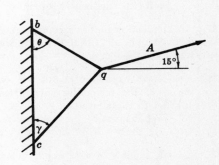

Fig. 10-40   **Fig. 10-41**

32. Wind strikes a barn at an angle of $70°$ with the side of the barn; it exerts a $3300\#$ total force on the barn in that direction. Find the parallel and perpendicular components of force with respect to the side of the barn. Use the vector scale $1'' = 1000\#$.
    Ans. parallel component $= 1130\#$, normal component $= 3100\#$

33. Force A of $1500\#$ acts on cables QB and QC as shown in Fig. 10-41 above. Springs record a $1050\#$ tension in cable QB and a $625\#$ tension in cable QC. Find the angles θ and γ. Use the vector scale $1'' = 300\#$.   *Ans.* $\theta = 85°$, $\gamma = 40°$

34. A 180# weight is suspended by the rope framework shown in Fig. 10-42 below. *BCE* is an equilateral triangle. Find the tensions in the cables *AB* and *CD*. Use the vector scale 1″ = 50#.
   Ans. *AB* = 132#, *CD* = 161#

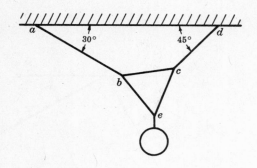

Fig. 10-42   Fig. 10-43

35. A steel beam weighing 1000# is shown in Fig. 10-43 above; its center of mass is at the center of the beam. Find the tensions in the supporting cables *A* and *B*. Use the vector scale 1″ = 200# and the distance scale 1″ = 2′.   *Ans.* *A* = 635#, *B* = 365#

36. Forces *A*, *B*, and *C* produce a torque on the bar shown in Fig. 10-44. *A* = 120#, *B* = 80#, *C* = 150#. Find (*a*) the single force which will keep the bar in equilibrium, and (*b*) its line of action. Use the vector scale 1″ = 50# and the distance scale 1″ = 2′.
   Ans. (*a*) Force = 168#, (*b*) Line of action passes through a point 4′-9″ directly to the left of the center of mass.

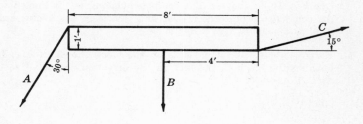

Fig. 10-44

37. An airplane registers a compass heading of N 120° W and an air speed of 450 m.p.h. The wind velocity is 100 m.p.h. at N 120° E. Find the ground speed and the actual course of the plane. Use the vector scale 1″ = 100 m.p.h.   *Ans.* Ground Speed = 410 m.p.h., Course = N 132° W

38. A rowboat, capable of a 5 m.p.h. velocity in still water, desires to reach a point six miles north across the river and five miles downstream. A water current, velocity of six m.p.h., is running due East, parallel to the shoreline.
       What must be the compass heading of the boat, and how long will it be until landing time?
   Ans. Heading = N 27° W, Time = 1.36 hours.

39. A boat, capable of a 12 m.p.h. velocity in still water, is travelling at maximum speed due East against a 5 m.p.h. water current flowing due West. The boat observes an island 1250 feet directly ahead, and decides to bypass the island to the North, missing it by 500 feet. Find the compass heading on which the pilot must set the boat. Use the vector scale 1″ = 5 m.p.h. and the distance scale 1″ = 250′.
   Ans. Heading = N 76° E

40. Boat *A*, capable of an 8 m.p.h. velocity in still water, is on an actual course directed due North. Wind blowing in the direction N 15° E tends to give the boat a 2 m.p.h. velocity vector in that direction. A 4 m.p.h. water current is flowing due West. Boat *B* is originally 9 miles West and 8 miles North of boat *A*, and travelling at 12 m.p.h. due East — actual course and speed with respect to land. 5 miles north of its original position, boat *A* sees boat *B*, and desires to intercept it.
       What is the closest that boat *A* can come to boat *B*? What must be its compass heading upon observing boat *B* in order to come this close? How much time will have elapsed in boat *A*'s travelling from its original position until the closest point is reached? Use the vector scale 1″ = 4 m.p.h. and the distance scale 1″ = 2 miles.   *Ans.* Heading = N 38° E, Distance = 0.9 mile, Time = 49 min.

41. A 180# plank (center of mass at the geometrical center of the plank) supports a 425# ball as shown in Fig. 10-45 below. Only the right end of the plank is firmly attached to a surface. Find the normal and tangential components of force at each supporting surface. Use the vector scale $1'' = 200\#$ and the distance scale $1'' = 2'$. (*Hint*: Find the resultant force acting on the plank at the right end.)
Ans. Inclined surface: normal comp. = 278#

Level surface: normal comp. = 348#,
tangential comp. = 103#

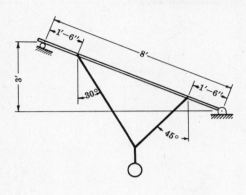

Fig. 10-45

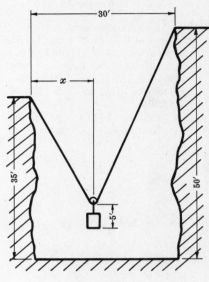

Fig. 10-46

42. A 70′ rope hangs over a gorge, supporting a 150# box hanging from a pulley. Refer to Fig. 10-46 above. Find the distance X when the box is closest to the ground. What is the tension in the rope at this point? How far above ground is the bottom of the box at this point? Use the vector scale $1'' = 50\#$ and the distance scale $1'' = 10'$.
Ans. Distance $X = 11'-6''$, Tension = 83#, Height = 5′–8″

43. A system of three forces ($A = 40\#$, $B = 24\#$, and $C = 35\#$) is shown in Fig. 10-47 below. When the system is in equilibrium position, it's found that the pulleys are at the same height, and separated by a distance of five feet. The boom is free to swing out from the wall. When the system is in equilibrium, find the distance X, the tension in the weightless boom, and the angle θ. Use the vector scale $1'' = 20\#$ and the distance scale $1'' = 1'-0''$.
Ans. Distance $X = 3'-6''$, Tension = 66.5#, $\theta = 18°30'$

44. Three guy ropes hold a pole upright as shown in Fig. 10-48. The pole feels a 100# compression. Find the stresses in the ropes. Use the vector scale $1'' = 50\#$.
Ans. $QA = 33\#$ tension, $QB = 42\#$ tension,
$QC = 59.5\#$ tension

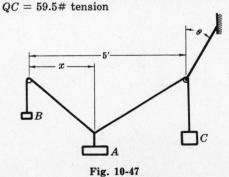

Fig. 10-47

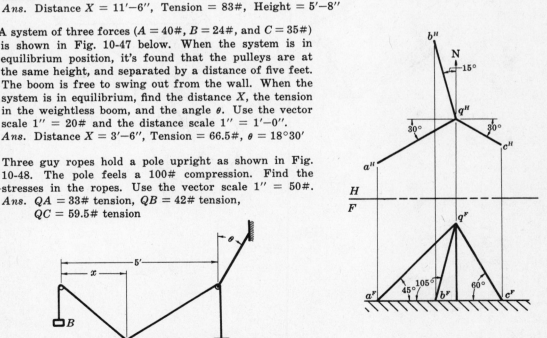

Fig. 10-48

45. Fig. 10-49 shows a framework supporting a 220# ball. Find the stress in each leg of the framework, using Bow's Notation. Use the vector scale $1'' = 100\#$.

   Ans. $QA = 115\#$ tension
   $QB = 200\#$ compression
   $QC = 85\#$ tension

46. Fig. 10-50 below shows one-half of the supporting system of a steel roller. QA is known to be the weakest cable, capable of standing a 450# tension. Find the heaviest roller that can be suspended, and the force this roller exerts against the slant surface. Use the vector scale $1'' = 200\#$.

   Ans. Roller = 1740#
   Normal force = 900#

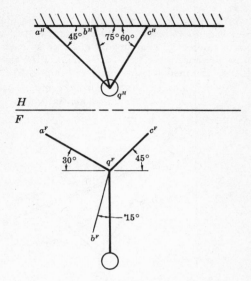

Fig. 10-49

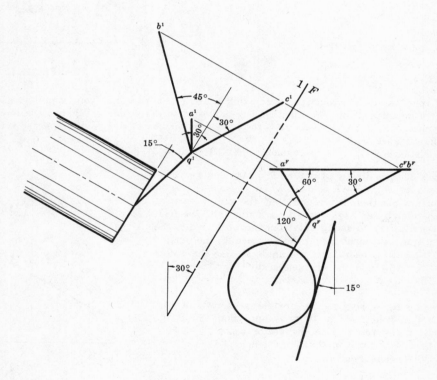

Fig. 10-50

APPENDIX

Frequently it will be found necessary to perform certain plane constructions in the solving of Descriptive Geometry problems. The most common ones are herein described and illustrated. In all probability, the student is familiar with these methods, but it has been the author's experience that the elementary processes often slip one's mind. Thus, these constructions are appended in order to provide the student with a handy reference source.

A.1 To DRAW a LINE PARALLEL or PERPENDICULAR to a GIVEN LINE

See Fig. A-1. Let XY be the given line. Using a T-square and a triangle with its hypotenuse along the T-square blade, adjust both instruments until one of the triangle legs corresponds to the line XY. Hold the T-square firmly and slide the triangle along the T-square to the proper place and draw the line VW parallel to XY using the same leg of the triangle. For a perpendicular line use the other leg of the triangle to draw MP perpendicular to XY.

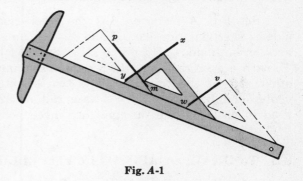

Fig. A-1

A.2 To DIVIDE a STRAIGHT LINE into any NUMBER of EQUAL PARTS

See Fig. A-2. Suppose it is required to divide the line XY into six equal parts. From either end of the line draw a line of indefinite length making any angle with XY; e.g., XZ. Using dividers, step off six equal segments on XZ and connect point 6 with point Y. From points 1, 2, 3, 4, and 5, construct lines parallel to line 6-Y. These lines will cut XY at the points 1', 2', 3', 4', and 5', which divide XY into the required six equal parts.

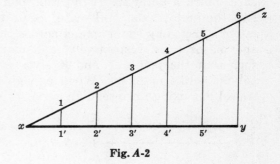

Fig. A-2

A.3 To DRAW the INSCRIBED CIRCLE of a TRIANGLE

See Fig. A-3 below. Let the given triangle be XYZ. Using an arbitrary radius and any of the vertices of the triangle as a center — e.g., Y — draw an arc which cuts the two sides of the triangle forming the angle. Using these two intersection points U and V as centers, draw two arcs of equal radius. Draw the line YW to the intersection point W of the two arcs, and extend the line past W. Similarly, construct the angle bisector of one other angle: the two bisectors will meet at a point T. From T, draw a line perpendicular to any side and draw the required circle, using this distance as a radius and T as a center.

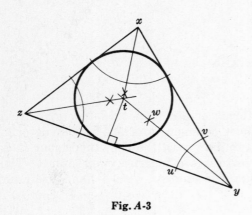

Fig. *A*-3

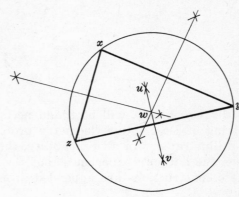

Fig. *A*-4

A.4 To DRAW the CIRCUMSCRIBED CIRCLE of a TRIANGLE

See Fig. *A*-4 above. Let the given triangle be *XYZ*. Using the end points of any side of the triangle — e.g., *Z* and *Y* — as centers, and a radius larger than half of *ZY*, pass arcs on both sides of the line *ZY*. Connect the two intersection points of the arcs, *U* and *V*, with a straight line. In a similar manner construct the perpendicular bisector of one other side of the triangle. The two bisectors will meet at a point *W*. Using *W* as a center and *WX, WY,* or *WZ* as a radius, draw the required circle. This method can be used to pass a circle through any three points not on the same straight line.

A.5 To DRAW an APPROXIMATE ELLIPSE

(*a*) The Trammel Method. See Fig. *A*-5(*a*) below. The major and minor axes of the ellipse must be known. Draw the two axes perpendicular to each other at their respective midpoints. Upon a stiff piece of paper with a straight edge, such as a 3″ × 5″ card, mark a point *X*. From this point mark another point *Y* such that *XY* is equal to half the major axis. Likewise, locate a second point *Z* such that *XZ* is equal to half the minor axis. Lay the trammel across the axes so that the points *Y* and *Z* touch the minor and major axes, respectively, and mark where the point *X* falls. Moving the trammel such that the points *Y* and *Z* always touch the two axes as shown, mark a number of points in like manner. When enough points have been thus located, the ellipse may be traced out using an irregular curve.

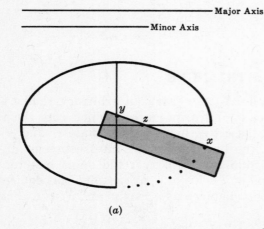

(*a*)

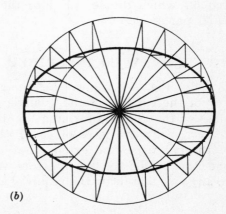

(*b*)

Fig. *A*-5

(b) **The Concentric Circle Method.** See Fig. *A*-5(*b*) above. Again the minor and major axes must be known. Draw the two axes perpendicular to each other at their respective midpoints, and draw two concentric circles from the point of intersection, using half the minor axis and half the major axis as radii. Draw a number of diameters of the larger circle at various angles with the major axis. From the points where these diameters intersect the smaller circle, draw lines parallel to the major axis. Similarly, from the points where the diameters meet the larger circle, draw lines parallel to the minor axis. The intersection point of the two lines drawn from the same diameter will be a point on the ellipse. Having a number of points thus located, draw the ellipse using an irregular curve.

A.6 To CONSTRUCT a REGULAR POLYGON in a GIVEN CIRCLE

See Fig. *A*-6 below. Draw a diameter of the circle and divide it into the same number of equal parts as the required polygon has sides, in this case 7. Construct an equilateral triangle, *OA*7′, on the diameter, and draw a line from the vertex of the triangle through the second point of division on the diameter, 2′. The distance *X* from point *A* to the intersection point *P* is now used to step off the vertices of the polygon on the circle.

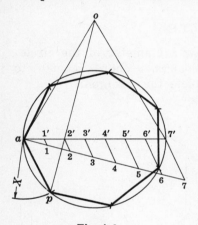

Fig. *A*-6

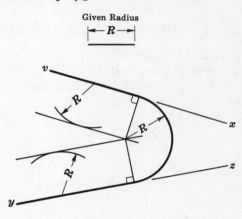

Fig. *A*-7

A.7 To DRAW an ARC of GIVEN RADIUS TANGENT to TWO STRAIGHT LINES

See Fig. *A*-7 above. Using the required given radius draw an arc from an arbitrarily selected point on each of the lines, and draw new lines parallel to them and tangent to the arcs. From the point of intersection of these new lines, draw perpendiculars to *VX* and *YZ* to locate the points of tangency. The required arc tangent to the two lines is then drawn between these points.

A.8 To DRAW a LINE TANGENT to a CIRCLE from an EXTERIOR POINT

See Fig. *A*-8. Using the T-square and triangle in conjunction as in Article *A*.1, move the triangle until one leg passes through the point *P* and is tangent to the circle. Slide the triangle along the T-square until the other leg passes through the center of the circle. This leg now intersects the circle at the point of tangency. Mark the point and draw the required tangent.

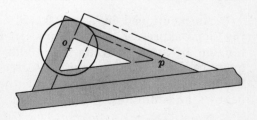

Fig. *A*-8

A.9 To RECTIFY the CIRCUMFERENCE of a CIRCLE

See Fig. A-9. Draw a line tangent to the circle at any point S. From point S, step off three times the diameter of the circle on the tangent line. Draw the diameter ST and the radius OR at an angle of 30° from the diameter ST. Construct RP perpendicular to ST and draw a line from C, the end of the tangent line, to point P. This distance is the circumference of the circle with an error of about 1/22,000.

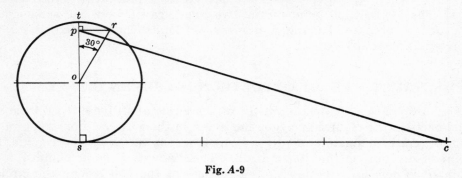

Fig. A-9

A.10 To CONSTRUCT a HEXAGON in a GIVEN CIRCLE

(*a*) **Triangle Method.** See Fig. A-10(*a*) below. Draw a diameter of the circle. At each end of the diameter, using the 30°-60° triangle, construct chords at 60° to the diameter and connect the free ends of the chords to complete the hexagon.

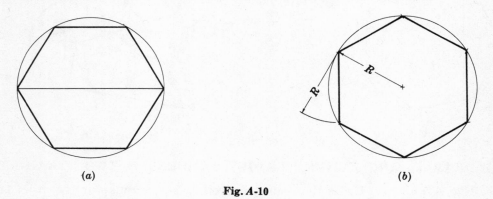

(a)

Fig. A-10

(b)

(*b*) **Divider Method.** See Fig. A-10(*b*) above. Using dividers set at a distance equal to the radius of the circle, step off successive arcs around the circle and connect these points in order to form the hexagon.

A.11 To MEASURE an ANGLE by its NATURAL TANGENT

As mentioned in Art. 2.3.1-*B*, the grade of a line is the slope expressed in per cent. In Fig. A-11 the natural tangent of angle ABC is CA divided by AB. Check the quotient obtained by consulting a table of natural tangents. The result will be the value of the angle expressed in degrees and minutes.

Note: The procedure is reversed if the per cent grade is required for a known angle.

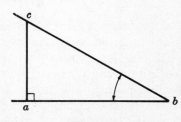

Fig. A-11

Bibliography of Standard Textbooks

There have been quite a number of excellent textbooks written on the subject of Descriptive Geometry. Often the approach used by one author is easier to understand than that of another. Therefore the following list of leading texts is presented so that the student can consult other books which adequately cover the major phases of Descriptive Geometry.

Bradley, H. C., and Uhler, E. H., *Descriptive Geometry For Engineers*, International Textbook Co.

French, Thomas E., and Vierck, Charles J., *Graphic Science*, McGraw-Hill Book Co.

Grant, Hiram E., *Practical Descriptive Geometry*, McGraw-Hill Book Co.

Higbee, F. G., *Drawing – Board Geometry*, John Wiley & Sons, Inc.

Hoelscher, Randolph P., and Springer, Clifford H., *Engineering Drawing and Geometry*, John Wiley & Sons, Inc.

Hood, George J., and Palmerlee, Albert S., *Geometry of Engineering Drawing*, McGraw-Hill Book Co.

Johnson, Lewis O., and Wladaver, Irwin, *Elements of Descriptive Geometry*, Prentice-Hall, Inc.

Paré, E. G., Loving, R. O., and Hill, I. L., *Descriptive Geometry*, Macmillan Co.

Rowe, Charles E., and McFarland, James D., *Engineering Descriptive Geometry*, D. Van Nostrand Co.

Schumann, Charles H., *Descriptive Geometry*, D. Van Nostrand Co.

Shupe, Hollie W., and Machovina, Paul E., *Engineering Geometry and Graphics*, McGraw-Hill Book Co.

Street, William E., *Technical Descriptive Geometry*, D. Van Nostrand Co.

Warner, Frank M., and McNeary, Matthew, *Applied Descriptive Geometry*, McGraw-Hill Book Co.

Watts, Earl F., and Rule, John T., *Descriptive Geometry*, Prentice-Hall, Inc.

Wellman, B. Leighton, *Technical Descriptive Geometry*, McGraw-Hill Book Co.

INDEX

SCHAUM'S INTERACTIVE OUTLINE SERIES

Schaum's Outlines and Mathcad™ Combined. . .
The Ultimate Solution.

NOW AVAILABLE! Electronic, interactive versions of engineering titles from the Schaum's Outline Series:

- *Electric Circuits*
- *Electromagnetics*
- *Feedback and Control Systems*
- *Thermodynamics For Engineers*
- *Fluid Mechanics and Hydraulics*

McGraw-Hill has joined with MathSoft, Inc., makers of Mathcad, the world's leading technical calculation software, to offer you interactive versions of popular engineering titles from the Schaum's Outline Series. Designed for students, educators, and technical professionals, the *Interactive Outlines* provide comprehensive on-screen access to theory and approximately 100 representative solved problems. Hyperlinked cross-references and an electronic search feature make it easy to find related topics. In each electronic outline, you will find all related text, diagrams and equations for a particular solved problem together on your computer screen. Every number, formula and graph is interactive, allowing you to easily experiment with the problem parameters, or adapt a problem to solve related problems. The *Interactive Outline* does all the calculating, graphing and unit analysis for you.

These "live" *Interactive Outlines* are designed to help you learn the subject matter and gain a more complete, more intuitive understanding of the concepts underlying the problems. They make your problem solving easier, with power to quickly do a wide range of technical calculations. All the formulas needed to solve the problem appear in real math notation, and use Mathcad's wide range of built in functions, units, and graphing features. This interactive format should make learning the subject matter easier, more effective and even fun.

For more information about *Schaum's Interactive Outlines* listed above and other titles in the series, please contact:

Schaum Division
McGraw-Hill, Inc.
1221 Avenue of the Americas
New York, New York 10020
Phone: 1-800-338-3987

To place an order, please mail the coupon below to the above address or call the 800 number.

--- ✂ --

Schaum's Interactive Outline Series
using Mathcad®

(Software requires 80386/80486 PC or compatibles, with Windows 3.1 or higher, 4 MB of RAM, 4 MB of hard disk space, and 3 1/2" disk drive.)

AUTHOR/TITLE	Interactive Software Only ($29.95 ea) ISBN	Quantity Ordered	Software and Printed Outline ($38.95 ea) ISBN	Quantity Ordered
MathSoft, Inc./DiStefano: Feedback & Control Systems	07-842708-8	_____	07-842709-6	_____
MathSoft, Inc./Edminister: Electric Circuits	07-842710-x	_____	07-842711-8	_____
MathSoft, Inc./Edminister: Electromagnetics	07-842712-6	_____	07-842713-4	_____
MathSoft, Inc./Giles: Fluid Mechanics & Hydraulics	07-842714-2	_____	07-842715-0	_____
MathSoft, Inc./Potter: Thermodynamics For Engineers	07-842716-9	_____	07-842717-7	_____

NAME_____ ADDRESS_____

CITY _____ STATE_____ ZIP_____

ENCLOSED IS ❑ A CHECK ❑ MASTERCARD ❑ VISA ❑ AMEX (✓ ONE)

ACCOUNT #_____ EXP. DATE _____

SIGNATURE_____

MAKE CHECKS PAYABLE TO McGRAW-HILL, INC. PLEASE INCLUDE LOCAL SALES TAX AND $1.25 SHIPPING/HANDLING

SCHAUM'S SOLVED PROBLEMS SERIES

- Learn the best strategies for solving tough problems in step-by-step detail
- Prepare effectively for exams and save time in doing homework problems
- Use the indexes to quickly locate the types of problems you need the most help solving
- Save these books for reference in other courses and even for your professional library

To order, please check the appropriate box(es) and complete the following coupon.

☐ **3000 SOLVED PROBLEMS IN BIOLOGY**
ORDER CODE 005022-8/**$16.95 406 pp.**

☐ **3000 SOLVED PROBLEMS IN CALCULUS**
ORDER CODE 041523-4/**$19.95 442 pp.**

☐ **3000 SOLVED PROBLEMS IN CHEMISTRY**
ORDER CODE 023684-4/**$20.95 624 pp.**

☐ **2500 SOLVED PROBLEMS IN COLLEGE ALGEBRA & TRIGONOMETRY**
ORDER CODE 055373-4/**$14.95 608 pp.**

☐ **2500 SOLVED PROBLEMS IN DIFFERENTIAL EQUATIONS**
ORDER CODE 007979-x/**$19.95 448 pp.**

☐ **2000 SOLVED PROBLEMS IN DISCRETE MATHEMATICS**
ORDER CODE 038031-7/**$16.95 412 pp.**

☐ **3000 SOLVED PROBLEMS IN ELECTRIC CIRCUITS**
ORDER CODE 045936-3/**$21.95 746 pp.**

☐ **2000 SOLVED PROBLEMS IN ELECTROMAGNETICS**
ORDER CODE 045902-9/**$18.95 480 pp.**

☐ **2000 SOLVED PROBLEMS IN ELECTRONICS**
ORDER CODE 010284-8/**$19.95 640 pp.**

☐ **2500 SOLVED PROBLEMS IN FLUID MECHANICS & HYDRAULICS**
ORDER CODE 019784-9/**$21.95 800 pp.**

☐ **1000 SOLVED PROBLEMS IN HEAT TRANSFER**
ORDER CODE 050204-8/**$19.95 750 pp.**

☐ **3000 SOLVED PROBLEMS IN LINEAR ALGEBRA**
ORDER CODE 038023-6/**$19.95 750 pp.**

☐ **2000 SOLVED PROBLEMS IN Mechanical Engineering THERMODYNAMICS**
ORDER CODE 037863-0/**$19.95 406 pp.**

☐ **2000 SOLVED PROBLEMS IN NUMERICAL ANALYSIS**
ORDER CODE 055233-9/**$20.95 704 pp.**

☐ **3000 SOLVED PROBLEMS IN ORGANIC CHEMISTRY**
ORDER CODE 056424-8/**$22.95 688 pp.**

☐ **2000 SOLVED PROBLEMS IN PHYSICAL CHEMISTRY**
ORDER CODE 041716-4/**$21.95 448 pp.**

☐ **3000 SOLVED PROBLEMS IN PHYSICS**
ORDER CODE 025734-5/**$20.95 752 pp.**

☐ **3000 SOLVED PROBLEMS IN PRECALCULUS**
ORDER CODE 055365-3/**$16.95 385 pp.**

☐ **800 SOLVED PROBLEMS IN VECTOR MECHANICS FOR ENGINEERS
Vol I: STATICS**
ORDER CODE 056582-1/**$20.95 800 pp.**

☐ **700 SOLVED PROBLEMS IN VECTOR MECHANICS FOR ENGINEERS
Vol II: DYNAMICS**
ORDER CODE 056687-9/**$20.95 672 pp.**

ASK FOR THE *SCHAUM'S* SOLVED PROBLEMS SERIES AT YOUR LOCAL BOOKSTORE OR CHECK THE APPROPRIATE BOX(ES) ON THE PRECEDING PAGE AND MAIL WITH THIS COUPON TO:

McGRAW-HILL, INC.
ORDER PROCESSING S-1
PRINCETON ROAD
HIGHTSTOWN, NJ 08520

OR CALL
1-800-338-3987

_____

NAME (PLEASE PRINT LEGIBLY OR TYPE)

_____

ADDRESS (NO P.O. BOXES)

_____

CITY   STATE   ZIP

ENCLOSED IS   ☐ A CHECK   ☐ MASTERCARD   ☐ VISA   ☐ AMEX   (✓ ONE)

ACCOUNT # _____   EXP. DATE _____

SIGNATURE _____

MAKE CHECKS PAYABLE TO MCGRAW-HILL, INC. <u>PLEASE INCLUDE LOCAL SALES TAX AND **$1.25** SHIPPING/HANDLING</u>. PRICES SUBJECT TO CHANGE WITHOUT NOTICE AND MAY VARY OUTSIDE THE U.S. FOR THIS INFORMATION, WRITE TO THE ADDRESS ABOVE OR CALL THE **800** NUMBER.